Expanding the Scope of End-Use Monitoring Policies to Address Operational Use of U.S.-Origin Weapons

AIDAN WINN, ALISA LAUFER, OMAR DANAF

Prepared for Office of the Secretary of Defense
Approved for public release; distribution is unlimited

For more information on this publication, visit **www.rand.org/t/RRA2961-2**.

About RAND

The RAND Corporation is a research organization that develops solutions to public policy challenges to help make communities throughout the world safer and more secure, healthier and more prosperous. RAND is nonprofit, nonpartisan, and committed to the public interest. To learn more about RAND, visit www.rand.org.

Research Integrity

Our mission to help improve policy and decisionmaking through research and analysis is enabled through our core values of quality and objectivity and our unwavering commitment to the highest level of integrity and ethical behavior. To help ensure our research and analysis are rigorous, objective, and nonpartisan, we subject our research publications to a robust and exacting quality-assurance process; avoid both the appearance and reality of financial and other conflicts of interest through staff training, project screening, and a policy of mandatory disclosure; and pursue transparency in our research engagements through our commitment to the open publication of our research findings and recommendations, disclosure of the source of funding of published research, and policies to ensure intellectual independence. For more information, visit www.rand.org/about/principles.

RAND's publications do not necessarily reflect the opinions of its research clients and sponsors.

Published by the RAND Corporation, Santa Monica, Calif.
© 2025 RAND Corporation
RAND® is a registered trademark.

Library of Congress Cataloging-in-Publication Data is available for this publication.

ISBN: 978-1-9774-1-4748

Cover image: Imago/Alamy

Limited Print and Electronic Distribution Rights

This publication and trademark(s) contained herein are protected by law. This representation of RAND intellectual property is provided for noncommercial use only. Unauthorized posting of this publication online is prohibited; linking directly to its webpage on rand.org is encouraged. Permission is required from RAND to reproduce, or reuse in another form, any of its research products for commercial purposes. For information on reprint and reuse permissions, please visit www.rand.org/pubs/permissions.

About This Report

The U.S. government is required to conduct end-use monitoring (EUM) to ensure that foreign end users of U.S. defense articles and services are complying with the requirements of government-to-government agreements with respect to use, transfers, and security. EUM is not currently structured or resourced to monitor partners' operational use of transferred articles. However, in light of U.S. Department of Defense (DoD)–heightened commitments to mitigate civilian harm in U.S. military operations and efforts to integrate these principles into security cooperation programs, greater visibility into partner operational use of U.S.-supplied weapons is required to ensure partner compliance. The scope and purpose of EUM deserves to be revisited.

In this report, we analyze the scope of existing EUM programs and identify mechanisms to better align U.S. policies and programs with broader national security guidance and policy commitments to enshrine human rights and civilian harm mitigation considerations in arms-transfer decisions. We envision an expanded scope for EUM that includes monitoring end users' operational use of U.S.-origin weapons and propose possible avenues for implementing operational EUM that complement DoD's broader, ongoing efforts to mitigate the risk of civilian harm.

The research reported here was completed in September 2024 and underwent security review with the sponsor and the Defense Office of Prepublication and Security Review before public release.

RAND National Security Research Division

This research was sponsored by the Office of the Under Secretary of Defense for Policy, Global Partnerships Office, and conducted within the International Security and Defense Policy Program of the RAND National Security Research Division (NSRD), which operates the National Defense Research Institute (NDRI), a federally funded research and development center sponsored by the Office of the Secretary of Defense, the Joint Staff, the Unified Combatant Commands, the Navy, the Marine Corps, the defense agencies, and the defense intelligence enterprise.

For more information on the RAND International Security and Defense Policy Program, see www.rand.org/nsrd/isdp or contact the director (contact information is provided on the webpage).

Acknowledgments

We greatly appreciate the opportunity to conduct this research provided by Deputy Assistant Secretary of Defense for Global Partnerships Christopher Mewett in the Office of the Under Secretary of Defense for Policy and his director for security cooperation, Scott Buchanan. We appreciate the work of Paul Brotzen, our action officer, in convening key stakeholders and providing guidance throughout this study. We are grateful to Deputy Assistant Director for Global Execution Earl Kirkley in the Office of International Operations at the Defense Security Cooperation Agency (DSCA) and Acting Deputy Assistant Director for Global Execution George Lumpkins in the Office of International Operations at DSCA for their helpful collaboration in sharing their perspectives and for facilitating introductions to the EUM community. We recognize the substantial contributions of many anonymous interviewees from the U.S. Departments of Defense and State and various nongovernmental organizations, all of whom shared unique perspectives, which were essential to exploring this topic. Thank you to our reviewers, Pauline Moore and Elias Yousif, who improved our draft report tremendously, particularly through suggesting ways to make it more accessible to a wide audience. Thank you to our editor, Drew Robison, who improved the accuracy and clarity of the report with great efficiency and attention to detail. Any errors or omissions remain the sole responsibility of the authors.

Summary

Issue

In this report, we analyze the scope of existing end-use monitoring (EUM) programs and identify mechanisms to better align U.S. policies and programs with broader national security guidance and policy commitments to enshrine human rights and civilian harm mitigation considerations in arms-transfer decisions. The Arms Export Control Act requires the U.S. government to conduct EUM to ensure that foreign end users of U.S. defense articles are complying with the requirements of government-to-government agreements with respect to use, transfers, and security. To comply with that law's requirement, the U.S. Department of Defense (DoD) established the Golden Sentry program, which seeks to verify a partner's compliance with end-use assurances that U.S.-origin articles and services be used for only the purposes for which they were furnished. DoD has made heightened commitments to mitigate civilian harm in U.S. military operations and to integrate these efforts into security cooperation programs through building foreign partners' capacity to mitigate the risk of civilian harm. However, EUM is not structured or resourced to monitor partners' operational use of transferred articles.

To fulfill DoD commitments regarding civilian harm mitigation and to fully implement broader U.S. policies that identify human rights concerns as central in arms transfers, a capability to monitor how U.S.-origin weapons are employed by foreign partners in an operational context would need to be developed, organized, and resourced distinctly from Golden Sentry. The report presents our analysis from the body of laws and policies relevant to this capability requirement, and we propose possible avenues for DoD to engage in the monitoring of partners' operational use of U.S.-origin weapons to complement the department's broader, ongoing efforts to mitigate the risk of civilian harm.

Approach

The research approach for this study involved semistructured interviews with key stakeholders in the EUM and civilian harm mitigation communities. This study considered a new mission for DoD that exists at the intersection of established responsibilities, including EUM policies, related policies on foreign military sales, and civilian harm mitigation and response responsibilities. Interview protocols were tailored to gain insights based on interviewees' roles and to explore the policy issues, processes, and gaps into which interviewees had direct visibility. The interviewees were provided with general questions in advance of the conversations. The research methodology also included review and analysis of relevant legal statutes, congressional materials (e.g., bills, official correspondence between members of Congress and senior government officials, Congressional Research Service reports), DoD issuances and memoranda, DoD Inspector General and U.S. Government Accountability Office audits, and annual EUM reports published through the Department of State's Directorate of Defense Trade Controls. The outcome of the research highlights considerations and challenges that will be encountered in a scope expansion of EUM that would monitor partners' operational use of defense articles obtained from the United States and mitigate civilian harm caused by foreign partners. Following this analysis, we suggest some concrete initial actions for DoD to begin cultivating internal capacities to expand the scope of EUM.

Key Findings and Recommendations

Finding: Department of Defense Terminology Related to the Golden Sentry End-Use Monitoring Program Has Created Confusion

The term *end use* has contributed to confusion regarding the purpose and scope of U.S. efforts to apply oversight to arms transfers. *End use*, in the context of weapon transfer policy, implies battlefield use and creates expectations that U.S. monitoring is designed to provide oversight and accountability regarding partners' employment of U.S.-origin defense articles and address compliance with international humanitarian law. Because updated policy guidance is absent, misperceptions regarding the objectives of the Golden Sentry EUM program are likely to remain.

Recommendation: Establish New Policy Guidance That Clarifies Terminology and Assigns Responsibilities for the Department of Defense's Monitoring of Partners' Employment of U.S.-Origin Defense Articles

We recommend that the Office of the Secretary of Defense (OSD) develop new policy guidance that defines responsibilities for monitoring and investigating partners' operational use of U.S.-origin defense articles to mitigate risks of U.S. arms being used in contravention of U.S. and international laws. These responsibilities should be clearly distinguished from existing roles associated with the implementation of the Golden Sentry EUM program.

Finding: Department of Defense Monitoring of Foreign Partners' Operations Will Require New Roles, the Application of a Broad Variety of Skills, and More Resources

Expanding the scope of EUM to develop a distinct program that monitors operational end use of U.S. weapons will involve new roles, processes, and skills and will require additional resources. Some functions are consistent with the existing roles of Security Cooperation Organizations, but other requirements will naturally fall beyond their mission and will need to be assigned to other parts of DoD. The methods involved in this new mission might include satellite imagery analysis, use of geolocation technology, forensic analysis, postblast analysis associated with explosive ordnance, human intelligence collection, investigative journalism, and eyewitness interviews. Some of these skills might already reside in parts of DoD but will need to be applied using new processes.

Recommendation: Adopt a Phased Approach to Resourcing a Program Scope Expansion

To support a broadened scope for EUM, we recommend that OSD take a phased approach to evaluating, requesting, and integrating new resources and personnel. A phased approach would permit OSD to develop a greater understanding of the skills and capabilities that are required to implement new objectives and use these data to inform resourcing plans. To apply a rigorous approach to determining the resources required to expand the programs' scope, DoD would need to gather initial data and validate the human capital and manpower requirements to inform longer-term planning. Several program design elements would significantly affect the required resources.

Finding: Clear Interagency Standards Will Be Important in Implementing a Broadened End-Use Monitoring Policy

DoD maintains evidentiary standards for investigating civilian harm caused by its own military operations but has not yet adopted standards for investigating harm caused by partners' use of U.S.-origin equipment. Similar to DoD, the Department of State has also indicated that it will track and corroborate reports of harm stemming from partners' use of U.S.-origin equipment. A lack of coordination on these processes and standards could impede information sharing across the interagency and might create confusion among external parties that wish to submit information in support of investigations.

Recommendation: Collaborate Across the Interagency to Establish a Centralized System for Collecting Public Reports of Harm by Partners Using U.S.-Origin Defense Articles

DoD should work with the U.S. interagency to jointly define clear and consistent standards for evidence when investigating reports that allies or partners have caused civilian harm using U.S.-origin weapons. These standards should be clearly communicated to the public.

Finding: Existing End-Use Monitoring Policy Designates Weapons According to the Risks Associated with Diversion, and a New Risk Paradigm Is Needed for Operational End-Use Monitoring

DoD designates transferred defense articles for either routine or enhanced EUM according to concerns related to technology security and potential diversion. The existing system does not account for each transfer's risk with respect to civilian harm.

Recommendation: Develop a New Designation System for Weapon Transfers According to the Risks for Civilian Harm

As part of an expanded program that monitors operational usage of U.S. arms transfers, OSD should work with the Department of State, Defense Security Cooperation Agency, and the civilian harm mitigation and response community, including nongovernmental organizations, to develop a new system for assessing and designating the level of risk that U.S.-origin weapons will pose to civilian populations and environments. This designation system should be integrated into arms-transfer policy decisions and reflected in the resources assigned to operational monitoring.

Contents

About This Report iii
Summary v
Figures and Tables xiii

CHAPTER 1
Introduction 1
   Purpose and Methodology 4
   Organization of Report 7

CHAPTER 2
How End-Use Monitoring Functions 9
   Roles and Responsibilities 10
   Resourcing 13
   "EUM Is Not a Wartime Operation" 14
   Implementation Challenges 17
   The Existing Scope of Golden Sentry 19
   Conclusion 21

CHAPTER 3
Civilian Harm and End-Use Monitoring: A New Policy Context 23
   Saudi Arabia, U.S. Weapons, and Civilian Harm 24
   The Use of Explosive Weapons in Populated Areas 27
   Civilian Harm Mitigation and Response Action Plan 28
   Conventional Arms Transfer Policy 30
   Civilian Harm Incident Response Guidance 31
   Department of Defense Instruction 3000.17, *Civilian Harm Mitigation
      and Response* 33
   National Security Memorandum 20 35
   Growing Dissatisfaction with End-Use Monitoring's Scope 39
   Proposed Legislation Reflects Desire for More Expansive End-Use
      Monitoring Program 41
   Conclusion 45

CHAPTER 4
How an Expanded End-Use Monitoring Policy Might Include
  Operational Monitoring..47
  New Skills for a New Mission 48
  Policy Challenges 51
  Legal Challenges 57
  Strategic Challenges 60
  Implementation Roles for the Department of Defense........................62
  Conclusion 68

CHAPTER 5
Recommendations 69
  Establish New Policy Guidance That Clarifies Terminology and
    Assigns Responsibilities for the Department of Defense's
    Monitoring of Partners' Employment of U.S.-Origin Defense
    Articles 70
  Adopt a Phased Approach to Resourcing a Program Scope Expansion ...73
  Collaborate Across the Interagency to Establish a Centralized System
    for Collecting Public Reports of Harm by Partners Using U.S.-
    Origin Defense Articles..76
  Develop a New Designation System for Weapon Transfers According
    to Their Risks of Causing Civilian Harm 80
  Concluding Thoughts 83

Abbreviations..87
References 89

Figures and Tables

Figures
- 5.1. End-Use Monitoring Umbrella Concept 71
- 5.2. Phased Approach to Resourcing a Program Scope Expansion... 74

Tables
- 1.1. End-Use Monitoring Scope Expansion Analysis Process 8
- 3.1. Policy Guidance Relevant to Arms Transfers and Human Rights.. 38
- 4.1. Golden Sentry vs. Operational End-Use Monitoring............. 48

CHAPTER 1

Introduction

U.S. strategic guidance highlights the increasingly important role that allies and partners play in achieving mutual security objectives. To meet today's most complex security challenges, the United States must find ways to deepen partnerships with those who share its interests and values and exercise force according to shared norms while holding these security partners accountable when they violate human rights or fail to comply with the requirements laid out in U.S. and international law. The transfer of defense articles to allies and partners has long been a core instrument of U.S. foreign policy and is not only designed to support allies and partners in confronting shared security threats but also regarded as an indispensable tool for strengthening relationships and promoting U.S. interests. In 1981, Andrew Pierre wrote in *Foreign Affairs*:

> Arms sales have become, more than ever before, a crucial dimension of world politics. They are now major strands in the warp and woof of international affairs. Arms sales are far more than an economic occurrence, a military relationship, or an arms control challenge—arms sales are foreign policy writ large.[1]

These words still ring true as the scale and significance of U.S. arms transfers have grown immensely since the 1990s. Because the character of war has become more complex, the policy considerations that have emerged regarding how these transfers should be governed have also increased in complexity.

---

[1] Andrew J. Pierre, "Arms Sales: The New Diplomacy," *Foreign Affairs*, Vol. 60, No. 1, Winter 1981/1982.

The Arms Export Control Act (AECA) requires the U.S. government to conduct end-use monitoring (EUM) to ensure that foreign end users of U.S. defense articles are complying with the requirements in government-to-government agreements with respect to use, transfers, and security.[2] EUM exists to ensure that U.S.-origin articles and services are used for only the purposes for which they were furnished and that their end use does not undermine U.S. interests or security objectives. It is important to highlight that the primary concern underlying the existing EUM is the physical security of the items. According to the U.S. Department of State (DoS), recipients must "maintain the security of any item with substantially the same degree of protection afforded to it by the U.S. government," including the security of the U.S. technology the item might contain.[3] DoS describes EUM as "an important component to protecting sensitive U.S. military and intelligence technologies and ensuring American warfighters retain their military edge over their adversaries."[4] Although the U.S. commitment to EUM is considerable—and unique among nations with significant defense exports—its scope is focused to address a specific risk. The purpose of EUM is to help protect U.S. technology from falling into the hands of adversaries. EUM guards against the possibility that unintended recipients will use these articles to undermine U.S. military advantage. EUM includes all actions to prevent misuse or unauthorized transfer of defense articles or services from title transfer until disposal. The type of defense article or service generally determines the level of monitoring required, according to the *Security Assistance Management Manual* (SAMM) issued by the Defense Security Cooperation Agency (DSCA).[5]

Questions regarding the purpose of EUM and its appropriate scope have emerged from Congress, nongovernmental organizations (NGOs), scholars, and the policy communities that are charged with implementing EUM pro-

---

[2] Public Law 90-629, Arms Export Control Act, as amended through Public Law 118-31, National Defense Authorization Act for Fiscal Year 2024, December 22, 2023.

[3] DoS, "End-Use Monitoring of U.S.-Origin Defense Articles," fact sheet, Bureau of Political-Military Affairs, January 20, 2021b.

[4] DoS, 2021b.

[5] DSCA, "End Use Monitoring," in *Security Assistance Management Manual*, 2024a.

grams. Furthermore, because the U.S. Department of Defense (DoD) has made heightened commitments to the protection of civilians in U.S. military operations, a new policy context is emerging for the exercise of the United States' oversight of its arms transfers. These objectives are reflected in DoD's commitments to establishing and resourcing civilian harm mitigation and response as a component of security cooperation programs, including the introduction of the concept of *tailored conditionality* in the 2022 Civilian Harm Mitigation and Response Action Plan (CHMR-AP).[6] The CHMR-AP creates new institutions and processes to strengthen DoD's ability to mitigate civilian harm during military operations. The CHMR-AP also lays out a long-term plan through which DoD will engage in learning to enhance assessments and investigations associated with instances of civilian harm and improve DoD's ability to respond. Increased accountability and transparency are two main principles that are embedded in the CHMR-AP. The implementation of these efforts is laid out through a phased approach, which acknowledges that it will take time for DoD to develop the capacity to carry out these new responsibilities.

Since the release of the CHMR-AP, world events have continued to place a spotlight on the plight of civilians in conflict settings, particularly in the two (very different) contexts presented by conflicts in Ukraine and Gaza. The scale, pace, and character of U.S. arms transfers have significant implications in these conflicts—and many other conflicts around the globe—which raises questions about the appropriate U.S. role in ensuring responsible use of transferred weapons. Many voices in Congress have expressed frustration and dissatisfaction with the absence of robust programs to ensure accountability over operational use of U.S. arms and have argued that the narrow scope of existing EUM approaches is inconsistent with U.S. values and U.S. policy commitments to reduce civilian harm.

---

[6] The term *tailored conditionality* appears in Objective 9 of the CHMR-AP and refers to a policy of setting expectations with partners that "security cooperation programs can be modified in response to partners' CHMR outcomes" (DoD, *Civilian Harm Mitigation and Response Action Plan [CHMR-AP]*, August 25, 2022b, p. 27). Although *tailored conditionality* is a new term and its implementation is nascent, U.S. policy is to assess a partner's willingness and ability to comply with international law and human rights when considering a proposed transfer of defense articles or defense services, which is reflected in Conventional Arms Transfer Policy.

In this report, we analyze the scope of existing EUM programs and propose that DoD undertake efforts to begin establishing a new capability to monitor a partner's operational use of U.S.-origin defense articles. We view the development of a capacity to monitor operational use of U.S.-origin defense articles as consistent with national security guidance and U.S. policy commitments to enshrine protecting human rights and mitigating civilian harm in arms-transfer decisions. We also propose several possible avenues for DoD to organize and resource efforts to strengthen U.S. oversight and increase accountability in the operational end use of U.S.-origin defense articles.

New mechanisms to address operational end use would represent a significant departure from the existing scope and practice of DoD's EUM efforts and will require the development of new capabilities and processes in DoD and the interagency. However, by leveraging other investments in civilian harm mitigation and response that are already underway in DoD and throughout the U.S. government and by working closely with communities that have expertise in these challenges, DoD can strengthen its ability hold partners accountable for the ways in which they employ U.S.-origin defense articles.

Purpose and Methodology

The objective of our report is to identify opportunities and mechanisms to better align EUM policy with DoD commitments to mitigate and respond to civilian harm and to adhere to fundamental U.S. values and principles of foreign policy in arms transfers. In this report, we seek to develop actionable recommendations to inform policy and resourcing decisions to address the requirements of an operational EUM policy. The question of EUM purpose and scope is of particular interest in our study.

Our central research questions are the following:

- What are the existing EUM requirements and implementation mechanisms?
- How do DoD commitments related to civilian harm mitigation create a new policy context for modifying the scope of EUM?

- What approaches could be taken to develop an operational EUM capability in DoD, and how would this differ from existing EUM efforts?
- What are some of the legal, policy, and strategic challenges of implementing a broadened approach to EUM that includes operational EUM?

Our research methodology involved semistructured interviews with key stakeholders in the EUM and civilian harm mitigation communities. Because our research was framed and resourced as a mini-study, a limited number of interviewees were selected according to their particular knowledge of and experience with the focus of inquiry.[7] Few individuals have deep knowledge of both EUM policy and practice *and* civilian harm investigations and their related bodies of knowledge. Therefore, we identified individuals from different communities who could inform different elements of the scope expansion analysis process. Through integrating these perspectives, we identified gaps to be addressed in the development of an expanded program.

Our interviews were tailored to gain insights into the roles and responsibilities of the interviewees and the policy dimensions, processes, and tasks that they had direct visibility into from their professional experiences. We used several different protocols to maximize the data-gathering potential of the small pool of interviewees in an effort to capture the most-relevant experience and insights. The interviewees were provided with questions in advance of the conversations and a brief background on the purpose and goals of the study. The interviews took place between October 2023 and August 2024; many were conducted virtually to access geographically dispersed interviewees whose involvement would have otherwise been pre-

---

[7] Our study mainly relied on *purposive* sampling (often also referred to as *judgmental*, *selective*, or *subjective* sampling), which is a nonprobability method used for qualitative research that involves identifying and selecting individuals or groups of individuals who are especially knowledgeable about or experienced with a specific topic or phenomenon. In addition to knowledge and experience, purposive sampling relies on the availability and willingness of interviewees to participate and their ability to communicate experiences and opinions in an articulate and reflective manner. For further discussion of purposive sampling, see John W. Cresswell and Vicki L. Plano Clark, *Designing and Conducting Mixed Method Research*, 2nd ed., SAGE Publications, 2010.

cluded because of budget limitations. Several U.S. officials in the Washington, D.C., area were interviewed in person. Our sponsor, the Office of the Under Secretary of Defense for Policy, was also helpful in this regard and included a wide variety of stakeholders in our interim briefings, which facilitated additional data collection.

One of the challenges we faced involved adapting research protocols to a rapidly evolving policy context. In particular, the release of "National Security Memorandum on Safeguards and Accountability with Respect to Transferred Defense Articles and Defense Services," National Security Memorandum 20 (NSM-20), in February 2024 significantly affected the policy context by introducing new requirements associated with international arms transfers.[8] Many of our interviews were undertaken prior to the release of this policy. Consequently, the issues that NSM-20 raised were not reflected in those earlier discussions.

The research methodology also included review and analysis of relevant legal statutes, congressional materials (e.g., bills, official correspondence between members of Congress and DoD leadership, Congressional Research Service reports), DoD issuances and memoranda, DoD Inspector General and U.S. Government Accountability Office (GAO) audits, and annual EUM reports published by DoS's Directorate of Defense Trade Controls. The goal of this document review was to inform an analytical process in which the existing practice of EUM is compared with changes in the policy context. This policy context is shaped by new commitments to mitigate civilian harm and real-world events, which are driving increased congressional scrutiny over the purpose and scope of EUM policy.

To analyze the opportunities for EUM scope expansion and the implications of such an expansion for resourcing, we conducted a gap analysis by comparing existing EUM programming with emerging demands for monitoring operational use. We began by identifying key policy documents that frame the legal requirements and policy objectives related to our study. In this first phase of our research, we reviewed documents that guide policy on arms transfers, including Conventional Arms Transfer (CAT) Policy and

---

[8] White House, "National Security Memorandum on Safeguards and Accountability with Respect to Transferred Defense Articles and Defense Services," National Security Memorandum, NSM-20, February 8, 2024.

the AECA.[9] We also analyzed policies that address civilian harm mitigation objectives, including the CHMR-AP and NSM-20, to explore the implications of these policies in the context of monitoring partner use of U.S.-supplied weapons.

We then conducted an analysis of the supply side of EUM. We examined the workforce and organizational structures, including personnel, funding streams, and other resources, that contribute to EUM programming.

We followed our supply analysis with an analysis of emerging demand for EUM scope expansion by reviewing congressional requests for information, proposed legislation, and press releases. We also interviewed representatives from the U.S. government and from the NGO sector who work on monitoring partners' use of U.S.-origin equipment to understand their experiences and perspectives regarding changing expectations for EUM since Golden Sentry was established by DoD. Merging our insights from the supply-and-demand analyses, we identified gaps in policy, resourcing, and relevant skills that would need to be bridged to expand EUM's existing scope. Table 1.1 summarizes each of the five steps of this approach and the types of documents that were used to inform different phases of the analysis process.

Organization of Report

Chapter 2 provides an overview of how DoD's EUM program functions. In this chapter, we characterize the purpose of DoD's existing EUM program, summarize the major roles and responsibilities, and highlight some of the challenges that implementors face, including efforts to conduct EUM in hostile environments. The information in this chapter provides a foundation for our subsequent analysis regarding the challenges inherent in expanding the program. In Chapter 3, we examine how a variety of U.S. government and DoD commitments to mitigate civilian harm shapes the policy context for EUM. In Chapter 4, we explore possibilities for an expanded EUM policy that includes operational EUM and highlight some considerations that DoD

---

[9] White House, "Memorandum on United States Conventional Arms Transfer Policy," National Security Memorandum, NSM-18, February 23, 2023

TABLE 1.1
End-Use Monitoring Scope Expansion Analysis Process

Analysis Phase	Description	Key Data Sources
Strategic direction	Identify strategic direction and priorities in a policy context.	• CHMR-AP • CAT Policy • AECA • NSM-20 • DoD *Law of War Manual*
Supply analysis	Review and analyze existing workforce organizational structure and resourcing mechanisms for EUM.	• Chapter 8 of the SAMM • DoDIs • Interviews with EUM community • Golden Sentry annual reports
Demand analysis	Characterize the demand for scope expansion and forecast associated requirements.	• Congressional RFIs and correspondence addressed to Secretaries of Defense and State • Proposed legislation • Press releases • Interviews with U.S. and NGO officials
Gap analysis	Identify gaps in policy, resourcing, personnel, and training between existing scope and approach and expanded scope indicated by demand analysis.	• GAO reports • DoD Inspector General reports • Interviews with U.S. and NGO officials
Solution analysis	Develop recommendations to address gaps through allocation of resources, technological investments, and policy and legislative action.	• All resources listed above

NOTE: DoDI = Department of Defense instruction; RFI = request for information.

must anticipate in efforts to broaden the program's scope. Chapter 5 presents our recommendations on the way forward, and we suggest some initial steps to implement when monitoring partners' use of U.S.-origin weapons.

CHAPTER 2

How End-Use Monitoring Functions

Since the Cold War, U.S. policy has placed a significant focus on the maintenance of military technical superiority as an element of national security. The goal of fielding and maintaining advanced military capabilities ahead of adversaries has been an important driver in the establishment of various export control policies and other policy measures undertaken to protect U.S. technology. EUM originally emerged from this context and was designed as a means of mitigating the technological security risks associated with arms transfers.

The legislative basis for existing EUM programs is derived from the Section 40A of the AECA.[1] The AECA requires the establishment of an EUM program to ensure that foreign end users are complying with the agreement to use defense articles only for the purposes approved in Section 503 of the Foreign Assistance Act of 1961, which states that defense articles and services can be transferred to countries and international organizations only for purposes of internal security, legitimate self-defense, civic action, or regional/collective arrangements.[2] EUM is the program through which DoD fulfills its "continual responsibility (from the time of transfer until eventual disposal)" to ensure that defense articles and services of U.S.-origin are being used for these purposes.[3] The requirements in the AECA

---

[1] U.S. Code, Title 22, Chapter 39, Section 2785, End-Use Monitoring of Defense Articles and Defense Services.

[2] Public Law 87-195, Foreign Assistance Act of 1961, as amended through Public Law 118-83, September 26, 2024.

[3] DSCA, "End Use Monitoring (EUM) Responsibilities in Support of the Department of Defense Golden Sentry EUM Program," DSCA Policy Memorandum No. 02-43, December 4, 2002.

have led to the establishment of two separate EUM programs: the Golden Sentry program, managed by DoD and responsible for EUM of government-to-government transfers of defense articles, and the Blue Lantern program, managed by DoS and responsible for EUM of arms sold via direct commercial sales.[4] The focus of this study was limited to the role of DoD implementing EUM.

Roles and Responsibilities

The director of DSCA has been delegated authority by the Secretary of Defense and Under Secretary of Defense for Policy to administer DoD's Golden Sentry EUM program. The Golden Sentry program is focused on the security of defense articles and seeks to ensure that, even after these articles are in the possession of a foreign partner, their storage is secure, they are not resold or diverted to third parties, and they are not repurposed in a way in which they were not intended to be used.

The following language contained in the legislation establishing this EUM program highlights the primary concerns underpinning the requirement:

> In carrying out the program established under subsection (a), the President shall ensure that the program (1) provides for the end-use verification of defense articles and defense services that incorporate sensitive technology, defense articles and defense services that are particularly vulnerable to diversion or other misuse, or defense articles or defense services whose diversion or other misuse could have significant consequences; and (2) prevents the diversion (through reverse engineering or other means) of technology incorporated in defense articles.[5]

When Golden Sentry was established in 2002, the memorandum stated the program's core objectives clearly: "Golden Sentry's objectives are to minimize security risks to the USG [U.S. government], our friends and allies,

---

[4] U.S. Code, Title 22, Chapter 39, Section 2785.

[5] U.S. Code, Title 22, Chapter 39, Section 2785.

and assure compliance with established technology control requirements."[6] Golden Sentry EUM does not monitor operational employment of weapons or seek to govern their usage through the application of conditions in the agreements made at the time of transfer. EUM monitors the storage and custody of the items. Analysts have underscored that existing EUM is fundamentally about technology security, not human security.[7]

Golden Sentry is broadly tasked with monitoring U.S. defense articles that are transferred to foreign parties to decrease the risk of their misuse and to ensure that sensitive technology control requirements are upheld. In support of this, Golden Sentry conducts two types of monitoring activities: (1) routine EUM (REUM) and (2) enhanced EUM (EEUM). REUM is the default EUM process required for all government-to-government transferred defense articles. All transferred defense articles that are not designated as requiring EEUM receive routine monitoring. To carry out REUM, uniformed Security Cooperation Organization (SCO) personnel are tasked with conducting Compliance Assessment Visits (CAVs)—or Virtual Compliance Assessments (VCAs) since 2021—to inspect the physical security, storage, and other compliance agreements listed in the terms of sale in applicable foreign military sales (FMS) agreements.[8] These visits are done in conjunction with other SCO responsibilities.[9] Data collected during REUM visits are to be entered in the Security Cooperation Information Portal (SCIP). As discerned from our interviews, the SCIP assists DSCA in tracking EUM inspections and compliance by providing a centralized database for DSCA to monitor Golden Sentry metrics. In addition, the SCIP provides a checklist of inspection requirements for REUM and EEUM checks. REUM inspections—in the form of CAVs or VCAs—are required to be exe-

---

[6] DSCA, 2002.

[7] Ari Tolany, Alexander Bertschi Wrigley, Annie Shiel, Elias Yousif, and Lauren Woods, "Demystifying End-Use Monitoring in U.S. Arms Exports," Center for Civilian in Conflict, Security Assistance Monitor, and Stimson Center, September 16, 2021.

[8] DSCA, "DSCA DoD Golden Sentry End-Use Monitoring (EUM) Virtual Compliance Assessments (VCA)," DSCA Policy Memorandum No. 21-64, October 13, 2021.

[9] U.S. official, interview with the authors, January 16, 2024.

cuted on a quarterly basis for transferred articles.[10] However, regulations specify that SCOs are required to conduct only a certain number of REUM inspections and do not delineate a quantity or proportion of transferred defense articles that must be inspected. Additionally, tracking information, such as serial numbers, is not required to be captured during REUM compliance assessments.[11]

Defense articles that are designated by Congress, military departments, implementing agencies, DoD, or DOS as requiring EEUM are subject to an expanded set of FMS, physical security, and compliance assessment inspection requirements.[12] The basis for the EEUM designation is a judgment regarding the sensitivity of the technology being transferred and what level of risk is created by its transfer. The specific risk being considered is assessed through a lens of technology security, not civilian harm mitigation. Many weapons available for international sale are highly destructive and yet are not designated for EEUM.

SCOs are generally required to visually inspect 100 percent of in-country EEUM-designated defense articles within one year from the last annual inspection by evaluating such criteria as accountability procedures and inventory records.[13] Additional requirements are formally listed within a letter of offer and acceptance for EEUM-designated articles.[14] Enhanced requirements include serial number verification, thorough physical security checks, and increased inspections but can also include case-specific requirements. These duties are carried out by SCO personnel during CAVs and VCAs.

Chapter 8 of the SAMM delineates the specific procedures, roles, and responsibilities of U.S. officials in their execution of the Golden Sentry program, including the factors that DSCA considers when determining CAV

---

[10] Defense Security Cooperation University, "End-Use Monitoring and Third-Party Transfers," in *Security Cooperation Management,* 43rd ed., 2023a, p. 18-2.

[11] U.S. official, interview with the authors, January 16, 2024.

[12] See Section 8.4 of DSCA, 2024a.

[13] See Sections 8.4.1.1 to 8.4.1.2 of DSCA, 2024a.

[14] DoD, Office of the Inspector General, "Management Advisory: DoD Review and Update of Defense Articles Requiring Enhanced End-Use Monitoring," DODIG-2023-074, May 19, 2023, pp. 2–3; DSCA, 2024a, Figure C8.F2.

priorities. The SAMM outlines procedures in support of EUM before, during, and after the transfer of defense articles or services. The manual is periodically updated to reflect evolving EUM policy and procedures and integrates policy memoranda that DSCA has issued.[15] The responsibilities outlined in the SAMM provide a snapshot of how DoD's EUM is designed to operate. The responsibilities of four key players for the execution of EUM under the Golden Sentry program are laid out in the SAMM: DSCA, military departments and implementing agencies, combatant commands (CCMDs), and SCO personnel (including defense attaché offices and relevant U.S. diplomatic missions). DSCA's management structure for Golden Sentry includes one EUM manager based in Washington, D.C., for each of the geographic CCMDs.

SCOs must report all potential unauthorized end use, including unauthorized access, unauthorized transfers, security violations, or known equipment losses to the CCMD, DSCA, and DoS. SCOs are directed by the SAMM to "be alert to, and report on, any indication that U.S. origin defense articles are being used for unauthorized purposes, are being tampered with or reverse engineered, or are accessible by persons who are not officers, employees, or agents of the recipient government."[16]

Resourcing

EUM is funded through FMS administrative funds, which are administrative surcharges collected by the U.S. government from the sale of U.S. defense articles and defense services to foreign partners through the FMS program. DSCA allocates FMS and foreign military financing (FMF) administrative funds to implementing agencies, geographic CCMDs, and defense agencies, to meet the various requirements for the management of FMS and FMF programs. The annual allocation process includes forecasting two-year demand through the FMS Forecast process for FMS case implementation, soliciting implementing agencies' budget requests and supporting docu-

---

[15] DSCA, *FY 2010 Congressional Budget Justification: Foreign Assistance, Title IV Supporting Information*, 2009, p. 23.

[16] DSCA, 2024a.

mentation, validating requests for funds, and finalizing the upcoming fiscal year (FY) budget.[17]

In FY 2022, DSCA employed six full-time civilian employees to administer the Golden Sentry program, and 484 uniformed SCO personnel performed duties in support of Golden Sentry.[18] In total, Golden Sentry received $2.08 million in funding—$1.8 million was allocated to DSCA and $277,300 was allocated to CCMDs.[19] Funds directed toward DSCA headquarters cover civilian salaries, travel, training, and contractor support. Funds allocated to CCMDs cover SCO travel for training and EEUM compliance assessments. In comparison, the United States conducted $43.07 billion worth of arms sales through the FMS program during FY 2022.[20] Accordingly, for every dollar spent on Golden Sentry during FY 2022, approximately $24,000 in sales were conducted through the FMS program. This information is provided to demonstrate that the resources dedicated to EUM are very limited when viewed in the context of the scale of U.S. arms transfers.

"EUM Is Not a Wartime Operation"

The existing approach to EUM has been challenging to implement in several contexts. Travel restrictions in place during the coronavirus disease 2019 (COVID-19) pandemic made it impossible to conduct in-person inspections in support of EUM for an extended period.[21] During this time, DSCA introduced a policy that allowed partners to self-report to adapt to pandemic conditions. However, a much larger challenge has emerged in the effort to conduct EUM on the vast amounts of U.S. defense articles sent to Ukraine

---

[17] See Section 14.1.2 of DSCA, "Security Assistance Forecasting, Planning, Programming, Budgeting, Execution, and Audit," in *Security Assistance Management Manual*, 2024c.

[18] DoD, *End-Use Monitoring of Defense Articles and Services Government-to-Government Services*, 2022a.

[19] DoD, 2022a.

[20] DSCA, "FY22 Security Cooperation Figures Announcement," press release, January 25, 2023.

[21] U.S. official, interview with the authors, January 16, 2024.

since February 2022. As of December 2024, the United States had provided more than $61.4 billion in military assistance through a variety of different security assistance authorities and programs.[22]

Because of the absence of U.S. personnel who were physically present in country, EUM approaches were significantly modified for the Ukrainian context. DoD was unable to directly observe the delivery and transfer of many sensitive defense articles, and Ukrainian officials were asked to self-report the status of such articles. DSCA quickly developed new guidance to address the challenging environment. Nonpermissive conditions were coupled with increasing congressional and public scrutiny because of the speed and scale of the weapon transfers. In December 2022, DSCA issued an update to the SAMM that provided guidance for a modified approach to EUM to be used in hostile environments, including the use of partner self-reporting, and explained the necessity for these measures as follows:

> The current EUM policy is designed to be executed in a peacetime environment where U.S. government officials conduct periodic site visits to verify accountability and ensure storage facilities meet physical security requirements. However, a combat/hostile environment significantly impedes the ability of U.S. government officials to conduct site visits where the potential for serious injury or death exists during execution of EUM activities. As a result, DSCA may request that the PN [partner nation] carry out the basic tenets of the DoD EUM Golden Sentry Program to provide reasonable assurance that U.S. defense articles are being stored, secured, and used in accordance with the terms and conditions of the relevant Letters of Offer and Acceptance (LOA) and equipment transfer agreements.[23]

In practice, EUM policy in hostile environments allows waivers in certain circumstances. SCOs are permitted to submit hostile environment waivers for items in areas that meet the threshold for hostile environment

---

[22] U.S. Department of State, "U.S. Security Cooperation with Ukraine," fact sheet, Bureau of Political-Military Affairs, December 2, 2024.

[23] DSCA, "Conducting End Use Monitoring in a Hostile Environment," DSCA Policy Memorandum No. 22-87, December 20, 2022.

classification.[24] A 2024 GAO report that explored the adequacy of the modified approach to EUM in Ukraine found that there were significant gaps in data.[25] GAO recommended that DoD improve the accuracy of defense article delivery data and undertake a broader evaluation of its EUM approach in Ukraine. EUM implementers have highlighted the scope of the challenge in Ukraine, as evidenced in our discussion with an interviewee:

> EUM is not a wartime operation. It is a peacetime operation. In Iraq and Afghanistan, we had troops on the ground, which is different than Ukraine. The fact that we do not have troops in Ukraine makes it difficult to do EUM because we do not want people to be put in harm's way. 100-percent accountability of defense articles is difficult in a combat zone.[26]

Although the scale and pace of security assistance provided to Ukraine makes that conflict's challenges somewhat unique, the situation clearly highlights some of the inherent limitations of an EUM program that is based on in-person inspections, focuses on physical security, and relies on permissive conditions to verify partner compliance. One interviewee explained that the number of items requiring EEUM in Ukraine went from 12,000 to 60,000 in a period of 12 months, but the resources and manpower to conduct the required EEUM did not keep pace.[27]

Research suggests that the risks of diversion are highest in contexts in which weapons are being used—most likely in some form of conflict.[28] U.S. officials have acknowledged that information from regions in conflict is

---

[24] U.S. official, interview with the authors, January 16, 2024.

[25] Chelsa A. Kenney, *Ukraine: DOD Should Improve Data for Both Defense Article Delivery and End-Use Monitoring*, U.S. Government Accountability Office, GAO-24-106289, March 13, 2024.

[26] U.S. official, interview with the authors, January 25, 2024.

[27] U.S. official, interview with the authors, January 25, 2024.

[28] Conflict Armament Research, "Typology of Diversion: A Statistical Analysis of Weapon Diversion Documented by Conflict Armament Research," *Diversion Digest*, No. 1, August 2018.

often the most important but also hardest to obtain.[29] However, EUM policy and practice is largely focused on physical security as a means of mitigating risk of diversion and is based on presumptions about the accessibility of the environment in which U.S. defense articles will be monitored. Golden Sentry is primarily designed to operate in a permissive, peacetime environment.

Implementation Challenges

SCOs face resourcing challenges that constrain their ability to carry out EUM activities in support of Golden Sentry at the operational level. SCO personnel are tasked with executing Golden Sentry requirements in addition to other security cooperation responsibilities. Although almost all SCOs are assessed as complying with their assigned EUM responsibilities, according to our interviews with Golden Sentry managers and geographic CCMD points of contact, SCO personnel often have inadequate time and resources to conduct the tasks associated with all Golden Sentry requirements. SCO offices are not adequately staffed and juggle a variety of other responsibilities. Often, EUM is relegated to a lower priority than other SCO duties. As one U.S. official said, "There are 2,000 other things that take precedence over EUM for SCOs."[30] Because SCO offices get new personnel every two years, the rapid turnover creates pressure and a steep learning curve for many SCOs in mastering Golden Sentry responsibilities. Because of these challenges, REUM is often missed entirely.[31] When discussing the challenges faced by SCOs assigned with EUM duties, one U.S. official characterized the situation as follows:

> A lot of the problems we see are not blatant disregard for policies, but many of these SCO positions are personnel being rotated, [with] lack of institutional knowledge because of one- to three-year rotations. New EUM managers get minimal training in the SCO course because there

---

[29] U.S. official, interview with the authors, October 20, 2023.

[30] U.S. official, interview with the authors, January 16, 2024.

[31] U.S. official, interview with the authors, January 16, 2024.

is a lot to learn; a lot of it is on-the-job training. If they arrive at an inopportune time, they will be more likely to be delinquent.[32]

Although some officials said that training and support from EUM managers were yielding "small, incremental gains" in some regions, this progress has been extremely difficult to sustain because of a high rate of turnover in SCO positions; one U.S. official said, "Gains that are made are not consolidated because expertise is not retained in SCO offices. There's a constant stream of new personnel coming in."[33] In some locations, the rotations are very short. For instance, in U.S. Central Command, some EUM personnel rotate as frequently as every six months and are only able to take online training courses as preparation.[34]

Partner compliance can also be a challenge in implementing Golden Sentry requirements. Even though some foreign partners consider their adherence to EUM requirements to be a source of pride and are eager to demonstrate their compliance during the course of EUM inspections, there are times when partners appear to resent this intrusion and make it difficult for U.S. officials to schedule visits.[35] Some foreign partners have made an effort to leverage the EUM check as a means to derive other practical benefits. For example, partners might request maintenance or repairs to accompany the EUM check and, if this cannot be offered by the SCO, might respond by making it difficult to inspect. According to a U.S. official, "in practice, we sometimes need to offer the partner something of value for them to 'make it easy' to inspect."[36] In remote locations of countries, even the weather can sometimes make it impossible for the SCOs to reach a particular location. EUM managers report that they must take multiple factors into consideration when working to boost the compliance rate of SCOs in meeting program requirements of ensuring compliance with transfer agreements.[37]

---

[32] U.S. official, interview with the authors, January 25, 2024.

[33] U.S. official, interview with the authors, January 25, 2024.

[34] U.S. official, interview with the authors, January 30, 2024.

[35] U.S. official, interview with the authors, January 16, 2024.

[36] U.S. official, interview with the authors, January 16, 2024.

[37] U.S. officials, interview with the authors, January 30, 2024.

The Existing Scope of Golden Sentry

In our discussions with EUM implementers, we explored the limits of the existing program's scope to characterize what it is able to achieve with its structure, responsibilities, and funding. One of the major messages that emerged in our interviews was that Golden Sentry does not and has never attempted to have direct visibility into partners' operational use of transferred articles.[38] Because SCOs generally do not function as intelligence assets, they are not in a position to observe operational usage. Golden Sentry's definition of *use* refers to the partner not repurposing weapons differently from how they were originally designed, but this definition is not focused on operational use in combat nor does it address the human rights implications or impacts on civilians.[39] The FY 2021 Golden Sentry EUM report prepared for Congress made this explicit:

> Post-delivery monitoring activities are conducted to verify the partner's continued ownership of U.S. government provided equipment and also the partner's compliance with any applicable technology control or physical security requirements. They are not intended to address a partner nation's operational use of transferred equipment.[40]

In one interview, an EUM implementer clarified that although Golden Sentry might take account of expended items that were used operationally, in the regular course of inventory checks, the question of whether that use was "appropriate" by U.S. standards is beyond the scope of EUM visits, and the partner does not have any responsibility to report operational usage in real time:

> They'll tell us that they expended items, but our inventories are based on an annual inventory.... Because we do not have insights into targets, we verify if they still have it in inventory or if it's expended. We [SCOs] are on an annual cycle for annual enhanced inventory, [and]

---

[38] U.S. official, interview with the authors, October 20, 2023.

[39] U.S. official, interview with the authors, October 20, 2023.

[40] DoD, *End-Use Monitoring of Defense Articles and Services Government-to-Government Services*, 2021.

there is no obligation for them to tell us immediately if they expended something if, in their view, it is for appropriate use.[41]

The structure of SCOs also imposes significant limits on the EUM mission. The sizes of the offices are based on assessment of the overall scope and scale of the security cooperation mission with that country, and personnel allocations are based on the scope of that relationship. Given how many other security cooperation responsibilities (e.g., FMS, FMF, training, excess defense articles) that SCOs undertake, SCOs are minimally staffed and generally have very little flexibility to augment staffing for EUM or provide surge support.

When we discussed the concept of expanding the scope of EUM with those involved in the implementation of Golden Sentry, some expressed concerns that the expanded scope would exceed the ability of the Golden Sentry workforce, explaining, "This is a 'big DoD' problem and cannot be solved by DSCA. Operational EUM would be a big DoD mission. The Golden Sentry program was designed for a peacetime environment, not a wartime one."[42] According to an interviewee, an operational mission would require formal documentation establishing roles among the relevant elements of DoD, DoS, and the intelligence community to develop the capability to effectively monitor operational usage. Another respondent drew attention to the issue of terminology and how the label "EUM" can end up conflating distinct missions and policy concerns:

> What I think people are really asking about is target verification and operational reporting. I think this needs to be defined because this takes on many different terms. We are trying to define lines instead of blurring them. . . . We owe clear definition to the community in general.[43]

This distinction was also underscored in the context of questions about how partner past performance can feed into future policy decisions about arms transfers. According to a U.S. official,

---

[41] U.S. official, interview with the authors, January 30, 2024.

[42] U.S. official, interview with the authors, January 30, 2024.

[43] U.S. official, interview with the authors, January 30, 2024.

when making FMS decisions to partners, one can take past history on EUM and operational reporting use (two separate criteria) to assess if they should sell. . . . But they [EUM compliance and operational use] are not the same. I'm not even saying they are related. They are two separate criteria that people in FMS area can use but operational—how do they use these weapons?—is different from EUM/Golden Sentry, which is focused on protecting technology and storage of devices and preventing diversion.[44]

Conclusion

The structure and resourcing of Golden Sentry reflects its limited scope. The Golden Sentry program is designed to oversee the monitoring of defense articles to ensure compliance with sensitive technology control requirements though inspecting whether physical security is adequate to prevent diversion. Although guidance has been developed to address the challenges of monitoring in hostile environments, the program largely operates on assumptions about peacetime access and permissive environments, which limits its effectiveness in environments in which active conflict is taking place. The concerns being raised about operational use of U.S.-origin weapons are arising in the context of armed conflicts, in environments in which U.S. access is limited or precluded. Even in carrying out the existing program, SCO personnel receive minimal training and face resourcing challenges, high personnel turnover, and competing responsibilities. These structural challenges affect the execution of EUM activities. Overall, even though the Golden Sentry program aims to ensure that transferred defense articles are used as intended and meet security requirements, its structure faces several challenges. Our overview of Golden Sentry's structure provides insights into the investments and changes that would be required to establish a program that has an expanded mandate to cover the operational EUM of transferred weapons.

---

[44] U.S. official, interview with the authors, January 30, 2024.

CHAPTER 3

Civilian Harm and End-Use Monitoring: A New Policy Context

A new policy context that makes the consideration of human rights more central in policy decisions regarding arms transfers has emerged. Although the policy developments discussed in this chapter represent a period of rapid evolution in the policy context, the U.S. policy focus on the civilian harm implications of arms transfers predates the CHMR-AP. In 2020, Heidi H. Grant, the director of DSCA at the time, stated,

> We're consistently working with our allies and partners to facilitate the employment and military capabilities consistent with our values. We have been particularly focused on minimizing civilian harm resulting from their operations including expanding training, ensuring the provision of targeting capabilities to partners, and providing additional advisory support, with a specific emphasis on mitigation of civilian harm.[1]

To explore the appropriate scope of DoD's EUM and determine which approaches might make it consistent with broader policy commitments, we must consider the recent landscape of policy change regarding civilian harm mitigation. Therefore, we begin this chapter with a brief background on the history of U.S. arms transfers to Saudi Arabia to illustrate how that relationship has contributed to changes in the policy context. Afterward, we highlight some of the key developments in which U.S. policy positions

---

[1] Heidi H. Grant as quoted in David Vergun, "Officials Describe How Arms Sales Benefit the U.S., Partners," U.S. Department of Defense, December 4, 2020.

on the responsibility to mitigate civilian harm are expressed. The ways in which U.S. commitments to mitigate civilian harm have been articulated suggest a need for both restraint in arms transfers and a mechanism to strengthen accountability in partners' operational end use. In this chapter, we provide context for the recommendations in Chapter 5 and help illustrate why a scope expansion to EUM is an appropriate policy avenue at this stage to make the architecture of U.S. policies related to arms transfers and civilian harm mitigation more consistent.

Saudi Arabia, U.S. Weapons, and Civilian Harm

The United States' arms-transfer relationship with Saudi Arabia has been a driver of increased scrutiny and debate over U.S. policies. Globally, Saudi Arabia is the largest purchaser of U.S. arms, and these purchases account for approximately 15 percent of overall U.S. arms exports.[2] The Saudi-led coalition began its military involvement in 2015 to support the Republic of Yemen in its civil war against multiple insurgent factions. The coalition's main military opponent has been the Iran-backed Houthi rebels, who controlled large swaths of territory in western Yemen. To fight the Houthis, the Saudi-led coalition launched a protracted air campaign, leading to large amounts of harm on Yemen's civilian population.[3] Civilian deaths resulting from coalition bombings are estimated to have reached 15,000 and have been accompanied by the large-scale destruction of civilian infrastructure, including hospitals, schools, and other public buildings.[4] In August 2020, the DoS Inspector General issued a report which found that the department had failed to take proper measures to reduce civilian deaths from U.S.-made

---

[2] Stockholm International Peace Research Institute, "Trends in International Arms Transfers, 2023," fact sheet, March 2024.

[3] Jeremy M. Sharp, *Yemen: Civil War and Regional Intervention*, Congressional Research Service, R43960, November 23, 2021.

[4] Armed Conflict Location and Event Data, "Country Hub: Yemen," webpage, undated.

precision-guided bombs transferred to Saudi Arabia and the United Arab Emirates and used in the Yemen war.[5]

The United States has transferred a wide array of aircraft, missiles, and small diameter bombs to Saudi Arabia since the Yemen intervention began, which has increased concerns that these defense articles might have been used to conduct these strikes.[6] As a result of the large amount of military assistance from the United States—a total of $54.6 billion between FY 2015 and FY 2021—concerns mounted regarding the possibility that human rights violations were committed with U.S.-supplied arms.[7]

One notable example that illustrates the accountability challenges in this relationship came in 2018, when Saudi Arabia admitted negligence in a strike on a school bus that directly resulted in the death of 40 children.[8] Analysis of the blast site by Bellingcat determined that the front control fin of a 500-pound GBU-12 Paveway II bomb was found near the blast site.[9] Photographs of a serial number listed on the control fin indicate that the munition was manufactured by a U.S. firm and was likely provided to Saudi Arabia in a U.S. transfer of 4,020 GBU-12s in 2015.[10] Several challenges have prevented the collection of specific operational information that could confirm the type of weapon used or the civilian death toll.

On February 4, 2021, the Biden administration announced that it was taking steps to "course-correct" U.S. foreign policy to better align it with democratic values and that one of those measures would include "ending all American support for offensive operations in the war in Yemen, includ-

---

[5] DoS, Office of the Inspector General, *Review of the Department of State's Role in Arms Transfers to the Kingdom of Saudi Arabia and the United Arab Emirates*, August 2020.

[6] Jason L. Blair, *Yemen: State and DOD Need Better Information on Civilian Impacts of U.S. Military Support to Saudi Arabia and the United Arab Emirates*, GAO, GAO-22-105988, June 2022, pp. 12–13.

[7] Blair, 2022.

[8] Salma Abdelaziz, Alla Eshchenko, and Joe Sterling, "Saudi-Led Coalition Admits 'Mistakes' Made in Deadly Bus Attack in Yemen," CNN, September 2, 2018.

[9] Michael Cruickshank, "A Saudi War-Crime in Yemen? Analysing the Dahyan Bombing," Bellingcat, August 18, 2018.

[10] Cruickshank, 2018.

Expanding the Scope of End-Use Monitoring Policies to Address Operational Use

ing relevant arms sales."[11] The decision resulted from years-long concerns about civilian casualties in Yemen, including those resulting from the Saudi-led coalition air campaign and a strong desire to help bring the war to an end. To investigate concerns about Saudi operations using U.S.-supplied weapons, GAO conducted an inquiry to determine whether the relevant U.S. entities in DoD and DoS had assessed whether EUM violations were committed by the Saudi-led coalition using U.S.-transferred arms. The findings of the inquiry were published in 2022.[12]

DoD and DoS explained in their response to the GAO report that pre-transfer FMS assurances do not obligate Saudi Arabia to share targeting data with the United States.[13] Without targeting data, it is difficult to determine whether a specific munition was used at a specific date and time to corroborate potential visual evidence.[14] DoS has not investigated these incidents as potential EUM violations. According to government officials, if the U.S. government had investigated this incident, it would not have been possible to determine whether the partner's strike constituted misuse without knowing the partner's tactics, techniques, and procedures.[15] U.S. Central Command officials cited a lack of mechanisms to compel the Saudi government to provide targeting data on strikes, because Saudi Arabia and other coalition members were considered sovereign countries at war.[16] One of GAO's recommendations for DoD was to develop guidance, in consultation with DoS, on how to implement DoD policy, including clarifying DoD roles and responsibilities, for reporting any indications that U.S.-origin defense articles were used in Yemen by Saudi Arabia or the United Arab Emirates

---

[11] White House, "Remarks by President Biden on America's Place in the World," February 4, 2021.

[12] Blair, 2022.

[13] Blair, 2022, pp. 28–29.

[14] Blair, 2022, pp. 29–30; New America, "The War in Yemen," webpage, undated; NGO expert on civilian harm investigations, interview with the authors, January 25, 2024.

[15] Blair, 2022.

[16] Notably, Saudi Arabia and coalition members have maintained an informal truce in the conflict in Yemen since April 2022 that effectively ended cross-border attacks between all parties.

against anyone other than legitimate military targets or for other unauthorized purposes.[17]

The U.S.-Saudi arms-transfer relationship provides an important context for the discussion of expanding EUM because it has generated significant criticism and increased scrutiny. The remaining portion of this chapter highlights some key developments in which U.S. policy positions on the responsibility to mitigate civilian harm are expressed. The significance of U.S. support for Saudi Arabia and the findings of the GAO report played a significant role in DoS's decision to develop the Civilian Harm Incident Response Guidance (CHIRG), which is discussed at greater length in this chapter.[18]

The Use of Explosive Weapons in Populated Areas

The use of explosive weapons with wide-area impacts in areas of concentrated populations has clearly demonstrated indiscriminate impact, regardless of the precision capabilities with which they might be equipped.[19] The urbanization of armed conflict has made the use of explosive weapons in densely populated areas increasingly common. This has profound humanitarian consequences, including high death tolls and destruction of critical civilian infrastructure.

In response to this trend, the United States and 82 other states, including most North Atlantic Treaty Organization allies, endorsed the Political Declaration on Strengthening the Protection of Civilians from the Humanitarian Consequences Arising from the Use of Explosive Weapons in Populated Areas (EWIPA) in November 2022.[20] By signing this declaration, the

---

[17] Blair, 2022.

[18] Blair, 2022.

[19] The term *explosive weapons* generally refers to any munition (e.g., bombs, artillery shells, grenades, missiles, landmines, improvised explosive devices) that is activated by detonation of a high-explosive substance. Explosive weapons project blast or fragmentation from the point of detonation.

[20] Political Declaration on Strengthening the Protection of Civilians from the Humanitarian Consequences Arising from the Use of Explosive Weapons in Populated Areas

United States stated its commitment to implementing policies and practices designed to limit civilian harm and damage to the civilian environment caused by the use of explosive weapons in populated areas. The declaration seeks to improve data collection and promote the sharing of lessons to inform policies regarding the investigation of harm to civilian and methods of determining accountability. According to the International Committee of the Red Cross, this declaration is the first of its kind that commits states to reviewing their military policies and restrict or refrain from the use of explosive weapons in populated areas in the interest of mitigating civilian harm.[21]

In April 2024, representatives from more than 90 countries met in Norway for the Oslo Conference on the Political Declaration, the first international follow-up meeting to review the EWIPA declaration's implementation. DoD officials remarked that the United States has been proud to endorse the EWIPA declaration and that its signatories can accomplish "critically important work to improve the protection of civilians in armed conflict."[22]

Civilian Harm Mitigation and Response Action Plan

DoD released its CHMR-AP in December 2022.[23] The release of the CHMR-AP followed a lengthy period of increasingly vocal criticism from NGOs and media regarding the inadequacies of DoD's civilian harm mitigation efforts in the context of not only criticism regarding a perceived lack of transparency in DoD's investigation and response to civilian harm resulting from U.S. military operations but also a lengthy period of introspection in DoD among those who had witnessed the civilian harm implications.

(EWIPA), signed at Dublin, Ireland, November 18, 2022.

[21] Simon Bagshaw, "The 2022 Political Declaration on the Use of Explosive Weapons in Populated Areas: A Tool for Protecting the Environment in Armed Conflict?" *International Review of the Red Cross*, No. 954, December 2023.

[22] Dan Stigall, "The EWIPA Declaration and U.S. Efforts to Minimize Civilian Harm," *Articles of War* blog, May 22, 2024.

[23] DoD, 2022b.

U.S. Defense Secretary Lloyd J. Austin III, in particular, was deeply committed to the goal of strengthening DoD's capability to mitigate civilian harm, a conviction that was shaped by his own personal experience as the commander of U.S. Central Command. Secretary Austin grappled with the challenge of trying to reduce the numbers of civilian casualties in that theater while trying to preserve flexibility for commanders in the field. The CHMR-AP responded to many of the findings captured in an independent assessment carried out by RAND and CNA that examined DoD's standards, processes, procedures, and policies related to civilian casualties resulting from U.S. military operations and concluded that "DoD is not adequately organized, structured, or resourced to sufficiently mitigate and respond to civilian-harm issues."[24]

The CHMR-AP lays out a series of significant commitments by DoD to mitigate and respond to civilian harm resulting from DoD operations, including the creation of new institutions and processes. In addition to mitigating civilian harm during operations, the plan aims to facilitate continued learning throughout DoD and elevate this priority. Assigning resources to build expertise in civilian harm mitigation clearly signals a significant commitment to a cultural and mindset shift regarding the inevitability of civilian casualties. The plan recognizes that the objectives it lays out are both "ambitious and necessary."[25] Although much of the CHMR-AP focuses on developing the internal capacity of DoD to mitigate civilian harm in U.S. military operations, it also addresses the need to reflect these principles and commitments as a component of security cooperation programs and through applying tailored conditionality, as appropriate, to promote the efforts of U.S. allies and partners in Objective 9.

---

[24] Michael J. McNerney, Gabrielle Tarini, Karen M. Sudkamp, Larry Lewis, Michelle Grisé, and Pauline Moore, *U.S. Department of Defense Civilian Casualty Policies and Procedures: An Independent Assessment*, RAND Corporation, RR-A418-1, 2022, p. xii.

[25] DoD, 2022b.

Conventional Arms Transfer Policy

Several months after the release of the CHMR-AP, DoS published the revised CAT Policy of 2023.[26] The CAT Policy, which is the primary policy framework designed to inform and adjudicate U.S. arms-transfer decisions, distinguished itself markedly from its previous iteration by focusing more explicitly on the human impact and civilian harm implications of arms transfers. Heightened emphasis was placed on complying with the law of armed conflict and avoiding transfers that would contribute to violations of human rights and international humanitarian law (IHL). The policy also specifically committed to the following: "Strengthen ally and partner capacity to respect their obligations under international law and reduce the risk of civilian harm, including through arms transfers, as well as appropriate tools, training, advising, and institutional capacity-building efforts."[27] The CAT Policy makes an explicit connection between legitimacy of arms transfers and the protection of civilians from harm, and the policy affirms that the recipient government of an arms transfer is responsible for complying with conditions of arms transfers involving obligations under international law, including those relating to human rights. The CAT Policy stipulates that the United States will engage in "appropriate monitoring as part of its effort aimed at ensuring transferred arms are used responsibly and in accordance with these conditions and obligations."[28] The reference to *appropriate monitoring* is noteworthy because it signals a commitment to undertake monitoring designed to address the use of weapons, which is outside the scope of the Golden Sentry program.

The CAT Policy recognizes that, when used in a way that violates U.S. values, U.S. arms transfers pose a threat to U.S. interests, as evident in the following text from the memorandum on the CAT Policy:

> United States foreign policy and national security objectives are best advanced by facilitating arms transfers to trusted actors who will use them responsibly and who share United States interests. This policy

---

[26] White House, 2023.

[27] White House, 2023.

[28] White House, 2023.

recognizes that, when not employed responsibly, defense materiel can be used to violate human rights and international humanitarian law, increase the risk of civilian harm, and otherwise damage United States interests.[29]

It is also noteworthy that the CAT Policy establishes a risk-based standard of "more likely than not" to guide decisionmaking policies regarding arms transfers.[30] This standard is a different threshold from previous conventional arms-transfer policies, which referred to actual knowledge that the arms transfer would lead to atrocities or human rights abuses. This new standard of "more likely than not" is relevant to the discussion of appropriate evidentiary standards to be adopted in a future operational EUM policy, which we discuss in Chapter 4.[31]

Civilian Harm Incident Response Guidance

DoS moved toward expanding the scope of EUM with the announcement of its CHIRG. The development of the CHIRG came in response to findings of the GAO report that highlighted the lack of policy guidance available for responding to reports that Saudi Arabia's use of U.S. defense articles had contributed to civilian harm.[32] The guidance, which is not yet publicly available as of this writing, directs all U.S. embassies to monitor and report on instances of civilian harm perpetrated by partner governments with U.S.-origin weapons.[33] According to media reporting, under the CHIRG process, "officials will examine and attempt to corroborate allegations of abuse that

---

[29] White House, 2023.

[30] In this context, "more likely than not" refers to the likelihood that civilian harm occurred and that U.S.-origin weapons were involved in the incident.

[31] See John Ramming Chappell and Ari Tolany, "Unpacking Biden's Conventional Arms Transfer Policy," Lawfare Institute and Brookings Institution, March 1, 2023. Chappell and Tolany acknowledge the subjective nature of the "more likely than not" standard and the challenges associated with its legal enforcement.

[32] Blair, 2022.

[33] Charles O. (Cob) Blaha, "The State Department's Civilian Harm Incident Response Guidance: How to Make a Good Thing Better," *Just Security* blog, July 18, 2024.

come from diplomatic and intelligence channels, the United Nations, media or civil society groups."[34] The process will be led by DoS, with support from DoD, the intelligence community, and other agencies.

The CHIRG is a joint effort between two bureaus in DoS: the Bureau of the Political-Military Affairs (PM), which focuses on the development security partnerships to advance U.S. national security objectives, and the Bureau of Democracy, Human Rights, and Labor, which is responsible for promoting global respect for the rule of law, democratic institutions, and human rights. The Directorate of Defense Trade Controls in the PM implements the International Traffic in Arms Regulations and the U.S. Munitions List. The Directorate of Defense Trade Controls is also responsible for adjudicating reports of end-use violations, including suspected third-party arms transfers. However, it is not clear whether CHIRG reports that are investigated and determined to constitute misuse will also be classified officially as end-use violations.

DoS officials involved in the implementation of the CHIRG attest that to date, the majority of reports received have been assessed to be "credible."[35] Although implementation remains nascent, DoS officials involved in the CHIRG have highlighted that the process is not primarily designed to be "punitive," because one of the goals is determining the cause of violations to agreements (e.g., Were there errors in targeting? Are capability limitations a factor, or does the partner seems to demonstrate insufficient political will?). U.S. policy outcomes will be tailored in light of these findings.[36] The CHIRG is also described as deliberative and not designed to be a rapid response capability.

The conflict in Gaza began shortly after the CHIRG was communicated to embassies worldwide as policy guidance, and, according to U.S. officials implementing it, the war is functioning as a stress test for the new guidelines because many of the institutional relationships required are still being

---

[34] Missy Ryan, "Biden Administration Will Track Civilian Deaths from U.S.-Supplied Arms," *Washington Post*, September 13, 2023.

[35] U.S. officials, interview with the authors, April 3, 2024.

[36] U.S. officials, interview with the authors, April 3, 2024.

developed.[37] However, U.S. officials report that the pressure to implement the guidelines with urgency has also strengthened interagency relationships, including those between DoS and DoD.[38] Establishing minimum standards for the submission of reports is a delicate balance because DoS would like to move toward a standardized threshold, but, at the same time, officials have expressed that they do not want to discourage the submission of reports that might have incomplete information.[39]

The CHIRG makes an important step toward addressing a long-standing gap in guidance regarding roles and responsibilities for responding to reports that U.S.-origin defense articles have contributed to civilian harm, but the lack of publicly available information regarding the CHIRG's implementation status makes it difficult to assess progress. At this stage, the total volume of incident reports being collected is not clear, so it is not possible to generate meaningful data to inform resourcing decisions. DoS has yet to publish any public information on the CHIRG process, including roles and responsibilities, the number of cases reported to date, or any official determination or policy responses that have been made. Former DoS officials have encouraged DoS to adopt greater transparency regarding the implementation of the CHIRG.[40]

Department of Defense Instruction 3000.17, *Civilian Harm Mitigation and Response*

DoD has adopted comparatively more transparency in the publication of a new DoDI that lays out the process by which civilian harm resulting from U.S. military operations will be conducted. In December 2023, the DoDI

---

[37] U.S. officials, interview with the authors, April 3, 2024.

[38] U.S. officials, interview with the authors, April 3, 2024.

[39] U.S. officials, interview with the authors, April 3, 2024.

[40] See Blaha, 2024; Blaha's comments in Forum on the Arms Trade, "Moving Beyond Rhetoric: Overcoming Challenges to Better U.S. Policy," video, June 4, 2024; and Josh Paul, "This Is Not the State Department I Know. That's Why I Left My Job," *Washington Post*, October 23, 2023. Paul was formerly a director in DoS's PM, and Blaha was a director of DoS's Office of Security and Human Rights from 2016 until he retired in 2023.

titled *Civilian Harm Mitigation and Response* (DoDI 3000.17) was issued to establish DoD's enduring policies, responsibilities, and procedures for mitigating and responding to civilian harm.[41] This DoDI serves as a means of operationalizing the commitments outlined in the CHMR-AP and, as its press release explains, "ensures operational commanders are supported with institutional resources, tools, and capabilities to effectively implement law of war protections of civilians, and to enable further steps to protect civilians and to respond appropriately when civilian harm occurs."[42]

Through this DoDI, DoD establishes standardized civilian harm assessment processes for DoD. The DoDI lays out a three-phase process of initial reviews, civilian harm assessments, and investigations and adopts the "more likely than not" standard that civilian harm occurred.[43] The DoDI is designed to support a consistent and standardized approach across DoD to more effectively assess the effects of U.S. military operations on civilians. Although the DoDI does provide guidance for civilian harm mitigation in "partnered operations" (see Section 3.3), it does not address U.S. policy for responding to reports of harm caused by partners using U.S.-supplied defense articles in the course of military operations.[44] The CHMR-AP's phased implementation plan assigns the task of developing interim policy guidance for roles, responsibilities, and procedures for responding to a report of allegation of civilian harm by allies and partners to the Under Secretary of Defense for Policy, but the process for implementing that policy, including new roles and responsibilities, has not yet been addressed in a formal DoD issuance.[45] However, DoDI 3000.17 does establish many responsibilities that would be relevant in a mission of operational EUM,

---

[41] DoD, *Civilian Harm Mitigation and Response*, Department of Defense Instruction 3000.17, December 21, 2023c.

[42] DoD, "DOD Announces Release of Department of Defense Instruction and Website on Civilian Harm Mitigation and Response," press release, December 21, 2023b.

[43] The "more likely than not" standard is sometimes referred to as a "preponderance of evidence." For further discussion of the standards of proof in the context of uncertainty, see Kevin M. Clermont, "Staying Faithful to the Standards of Proof," *Cornell Law Review*, Vol. 104, No. 6, September 2019.

[44] DoD, 2023b, p. 27.

[45] DoD, 2022b, p. 28.

even though they are not explicitly linked with this purpose. We discuss these operational EUM responsibilities at greater length in Chapter 4.

National Security Memorandum 20

One year after the release of the new CAT Policy, the Biden administration published the "National Security Memorandum on Safeguards and Accountability with Respect to Transferred Defense Articles and Defense Services," NSM-20, in February 2024. Consistent with the principles introduced in the previous year's CAT Policy, NSM-20 went further in establishing new requirements for both DoD and DoS. Specifically, NSM-20 introduced the requirement of obtaining "credible and reliable written assurances" from arms-transfer recipients that the usage of transferred articles will be undertaken in accordance with IHL and, as applicable, other international law.[46] The policy stipulated that recipients must also commit not to impede the delivery of humanitarian assistance when such weapons are being used in armed conflict. The memorandum gave DoD and DoS 90 days from its issuance to report to Congress with an assessment of whether the use of U.S.-supplied articles articles had been found to be inconsistent with these standards and established best practices for mitigating civilian harm.

Although NSM-20 did not technically introduce any new legal requirements, it did place a new emphasis on existing provisions in the AECA and Section 620I of the Foreign Assistance Act of 1961 and referred to these existing laws when directing a new type of oversight requirement for U.S. arms transfers. An implementation letter addressed to both departments regarding NSM-20 provides additional context for understanding its intended purpose and makes clear that the absence of operational end-use oversight had become a growing concern for Congress.[47] The letter explicitly connects EUM with the new requirements of the memorandum, framing NSM-20's

---

[46] White House, 2024.

[47] Chris Van Hollen, Brian Schatz, Martin Heinrich, et al., National Security Memorandum on Safeguards and Accountability with Respect to Transferred Defense Articles and Defense Services implementation, letter to Antony Blinken, Lloyd J. Austin III, and Avril Haines, U.S. Senate, March 8, 2024.

requirements as a means to address perceived gaps in the program's scope by asking: "As part of current end-use monitoring programs, to what extent does the United States government track where and how U.S. defense articles are being used in current conflict areas, including in Ukraine, Israel, Gaza, the West Bank, Lebanon?"[48] The letter also requested information on how both agencies planned to assess and determine whether U.S. defense articles are in violation of international law or in contravention with best practices for mitigating civilian harm.[49]

A closer look at the origins of NSM-20 provides important context for understanding how this memorandum reflects both a policy objective and a demand signal for legislation. NSM-20 arose during debates regarding the Israel and Ukraine supplemental appropriations. During that period, Senator Chris Van Hollen pressed for a mechanism to build considerations of international law compliance into this assistance package.[50] He expressed the following views in a *Washington Post* op-ed, outlining the conditions to be applied to recipients of U.S. military assistance:

> The president has asked Congress for an additional $106 billion for national security, including assistance for Israel, Ukraine and some of our Indo-Pacific allies. As the Senate takes up this request, I'm working with a group of my colleagues on an amendment that would require that the weapons received by any country under this bill are used in accordance with U.S. law, international humanitarian law and the law of armed conflict.[51]

Van Hollen explained that the amendment would also require the President to report to Congress on whether countries receiving military equipment paid for by U.S. taxpayers were compliant with those requirements and whether the use of U.S.-supplied weapons was consistent with established

---

[48] Van Hollen et al., 2024.

[49] Van Hollen et al., 2024.

[50] Legislative expert on U.S. arms-transfer policies, interview with the authors, August 26, 2024.

[51] Chris Van Hollen, "Israel's War Against Hamas Is Just, but It Must Be Fought Justly," *Washington Post*, December 6, 2023.

presidential directives on arms transfers and DoD policies for mitigating civilian harm.

At the time, there was no bill to which the amendment could be attached, but the initiative gained the support of 18 additional senators.[52] According to an interview with a legislative expert closely following this process, concerns in the Biden administration about growing opposition in Congress helped to pave the way for a compromise to forestall further legislative action. Many of Van Hollen's concerns were reflected in NSM-20.[53] This context also helps to account for why the policy applies to only U.S. grant assistance and not FMS because the debate from which the policy arose centered on conditions to be attached to U.S. military assistance.

Table 3.1 reflects the evolution of the policy context in which EUM is being implemented. The table summarizes the key documents that together create a policy context in which the absence of operational EUM capacity appears inconsistent with both U.S. policy and the growing demand from Congress for greater oversight and accountability in the management of U.S. arms transfers. To implement an EUM program that is more consistent with other policies regarding civilian harm mitigation, DoD will have to modify its approach to EUM by developing new capabilities. Before discussing steps DoD can take to better align its programming with the new policy context in Chapter 4, we close this chapter with a discussion of growing dissatisfaction with EUM's scope among policymakers that could further shape the policy environment in the future.

---

[52] U.S. Senate, Safeguards and Accountability with Respect to Transferred Defense Articles and Defense Services, Amendment 1389 to H.R. 815, submitted February 7, 2024. A bill is subject to amendment as soon as the Senate begins to consider it. See Christopher M. Davis, *The Amending Process in the Senate*, Congressional Research Service, 98-853, September 16, 2015.

[53] Legislative expert on U.S. arms-transfer policies, interview with the authors, August 26, 2024.

TABLE 3.1
Policy Guidance Relevant to Arms Transfers and Human Rights

Document	Category	Purpose
Political Declaration on Strengthening the Protection of Civilians from the Humanitarian Consequences Arising from the Use of Explosive Weapons in Populated Areas (2022)	International political agreement	Acknowledges the risks to civilians associated with the employment of explosive weapons in urban areas and commits signatories to taking action to reduce harm to civilians resulting from such uses of explosive weapons
Civilian Harm Mitigation and Response Action Plan (CHMR-AP) (DoD, 2022b)	Agency-level guidance	Establishes protocols and assigns responsibilities for improving DoD's civilian harm mitigation policies and practices, including those related to arms transfers
National Security Memorandum 18, "Memorandum on United States Conventional Arms Transfer Policy" (White House, 2023)	U.S. government policy	Describes principles and guidelines governing arms transfers to allies and partners; prohibits arms transfers when it is "more likely than not" that the recipient will use the article(s) to commit certain violations of IHL
Civilian Harm Incident Response Guidance	Agency-level guidance	Describes an interagency process for evaluating and responding to instances of civilian harm caused by U.S. allies and partners using U.S.-origin defense articles
Department of Defense Instruction 3000.17, *Civilian Harm Mitigation and Response* (DoD, 2023)	Agency-level guidance	Establishes DoD policies, responsibilities, and procedures for mitigating and responding to civilian harm resulting from U.S. military operations, including partnered operations
National Security Memorandum 20, "National Security Memorandum on Safeguards and Accountability with Respect to Transferred Defense Articles and Defense Services," (White House, 2024)	U.S. government policy	Requires the Secretary of State to obtain written assurances from recipient governments of U.S. arms that U.S.-origin equipment will be used in accordance with IHL and other international law

Growing Dissatisfaction with End-Use Monitoring's Scope

Because commitments to protect civilians and mitigate harm have become more prominent in U.S. government policy, the narrow scope of EUM has been criticized increasingly by elements of Congress as being inconsistent with these broader commitments. Although DoD officials have made efforts to clarify the existing scope and limits of EUM for congressional audiences and explain that the Golden Sentry program does not address operational EUM, a growing number of senators have raised questions about why the U.S. government lacks the capacity to conduct any level of operational monitoring of transferred defense articles. Increasingly, many senators have expressed the conviction that the policy areas of weapon transfers and human rights promotion must be treated as inherently connected and that the inability to hold partners accountable for use of weapons in a manner inconsistent with international law has become politically intolerable.

In May 2023, Senators Elizabeth Warren, Bernie Sanders, and Mike Lee wrote to the Secretaries of Defense and State to state their dissatisfaction regarding the absence of policy oversight on the operational use of transferred articles, despite a recognition of broader progress on civilian harm represented by the CHMR-AP. They expressed their concerns as follows:

> Despite this progress, we continue to have serious concerns about how DoD and the State Department track and monitor U.S.-origin weapons. The failure of DoD and State to properly do so stands in contradiction to U.S. values and our nation's efforts to prevent and mitigate civilian harm. We have identified several areas of concerns, which we urge DoD and State to further evaluate and make concerted efforts to improve its end-use monitoring (EUM) policies.[54]

The Hamas terrorist attacks on Israel in October 2023 and Israel's military response added a more urgent dimension to discussions regarding the appropriate standards for the use of U.S. military aid by foreign partners. In

---

[54] Elizabeth Warren, Bernard Sanders, and Michael S. Lee, weapon use monitoring and civilian harm, letter to Joseph R. Biden, U.S. Senate, May 14, 2023.

December, several senators wrote to President Joe Biden, Secretary Austin, and Secretary Antony Blinken and demanded that weapons being supplied to Israel be used in a manner consistent with the protection of civilians, citing the CHMR-AP, the CHIRG, and EWIPA and arguing that weapons being transferred to Israel were being done so without appropriate conditions regarding their use: "Your administration must ensure that existing guidance and standards are being used to evaluate the reports of Israel using U.S. weapons in attacks that harm civilians in order to more rigorously protect civilian safety during Israel's operations in Gaza."[55]

In March 2024, at a hearing of U.S. Senate Armed Services Committee, Senator Warren questioned General Michael E. Kurilla, Commander of U.S. Central Command, about DoD's role in collecting data that would convey a more complete picture of whether U.S. arms transfers contribute to civilian harm.[56] Senator Warren expressed dissatisfaction with the lack of data available to inform decisionmaking and posed the question, "General Kurilla, can we get an accurate picture of whether U.S. arms transfers contribute to civilian harm if we don't collect information about how these weapons are actually used?"[57] Senator Warren acknowledged that that the data might be difficult to collect but cited DoS's commitment to collecting better data through the CHIRG and requested that DoD make a similar commitment.

---

[55] Elizabeth Warren, Jeffrey A. Merkley, Bernard Sanders, Tim Kaine, and Martin Heinrich, civilian harm from Israel's use of U.S. weapons in Gaza, letter to Joseph R. Biden, U.S. Senate, December 5, 2023.

[56] U.S. Central Command, "Senate Armed Services Committee Hearing Posture of United States Central Command and United States Africa Command in Review of the Defense Authorization Request for Fiscal Year 2025 and the Future Years," transcript, March 8, 2024.

[57] U.S. Central Command, 2024.

Proposed Legislation Reflects Desire for More Expansive End-Use Monitoring Program

Several legislative initiatives in 2022 and 2023 started to explicitly connect U.S. arms transfers with broader U.S. and DoD commitments to mitigate civilian harm. These efforts reflect a broadening sentiment that U.S. arms transfers should be conditioned upon agreements by partners to abide by international law, IHL, and law of armed conflict in their use of these weapons and that mechanisms to ensure compliance with these conditions must become part of a more comprehensive approach to EUM.

Values in Arms Export Act of 2022

In 2022, Senator Patty Murray proposed the Values in Arms Export Act of 2022, which would amend the AECA to define the specific principles of responsible behavior and compliance with human rights and law of war that the United States demands of the states that purchase U.S. weapons. To enforce these principles, the bill creates a mechanism whereby the executive branch, Congress, or a newly created "Human Rights and Law of War Oversight Board" can each designate a *country of concern* if the country appears to be violating those principles. A designation would last for three years, and the country would be required to submit to a program of enhanced monitoring. The monitoring program, administered by DoD, would require oversight of any use of U.S.-provided weapons. That information, along with any other relevant information (e.g., NGO reports), would then be provided to DoS, and the Secretary of State would determine whether any actions violated the specified principles. Per the bill, the oversight mechanism would involve "the observation and evaluation by members of the United States Armed Forces of the targeting process used by the country of concern, the employment of the acquired defense articles by the country of concern, the return of any unused defense articles, and the post-use assessment of damage and casualties," including direct observation of real-time video feeds or data and other appropriate sources of information that can "be independently authenticated by the United States Government."[58]

Safeguarding Human Rights in Arms Exports Act of 2023

In March 2023, House Foreign Affairs Committee Ranking Member Gregory W. Meeks and then–Senate Foreign Relations Committee Chair Bob Menendez, along with a group of Democratic lawmakers, introduced the Safeguarding Human Rights in Arms Exports Act of 2023. The bicameral legislation was designed to strengthen Congress's oversight role in arms transfers and was aimed at preventing U.S. arms from being implicated in human rights abuses.[59] The act proposed to reform the AECA to guarantee that the protection and promotion of human rights are legally enshrined as integral considerations of the export of arms and defense services to foreign countries. The bill proposed amendments to Section 40A(a)(2)(B), End-Use Monitoring of Defense Articles and Defense Services, of the AECA by adding a new clause to specify that articles transferred must not be used to violate IHL or international human rights law and that the Secretary of State shall report to the appropriate congressional committees on the measures that will be taken, including any additional resources needed, to conduct an effective EUM program to fulfill the requirement in the proposed clause.[60]

---

[58] U.S. Senate, Values in Arms Export Act of 2022, S. 3558, February 2, 2022. See also Office of U.S. Senator Patty Murray, "Values in Arms Export Act of 2022," February 2022.

[59] See U.S. Senate, Safeguarding Human Rights in Arms Exports Act of 2023 or the SAFEGUARD Act of 2023, S.1025, March 29, 2023; and U.S. House of Representatives, Safeguarding Human Rights in Arms Exports Act of 2023 or the SAFEGUARD Act of 2023, H.R. 1801, March 27, 2023. The Senate version was cosponsored by Senators Dianne Feinstein, Tim Kaine, Patty Murray, and Brian Schatz. The House of Representatives version was cosponsored by Representatives Susan Wild, Abigail Spanberger, Sara Jacobs, Joaquin Castro, and Dean Phillips.

[60] See Section 8, End-Use Monitoring of Misuse of Arms in Human Rights Abuses, in U.S. Senate, 2023.

Safeguards and Accountability with Respect to Transferred Defense Articles and Defense Services

In December 2023, several senators announced an amendment to require that the weapons received by any country under the proposed national security supplemental be used in accordance with U.S. law, IHL, and the law of armed conflict. The Safeguards and Accountability with Respect to Transferred Defense Articles and Defense Services amendment proposed that the President report to Congress on whether countries receiving U.S. military equipment were adhering to standards established in presidential directives on arms transfers and in DoD policies for reducing harm to civilians.[61] In addressing the purpose of the amendment, Senator Kaine stated,

> U.S. assistance to our allies and partners has always come with the expectation that they will follow international laws of war. This global amendment reaffirms the need to protect innocent civilians caught in conflict zones and ensure the delivery of humanitarian aid to vulnerable populations. We must ensure that even as we stand with our partners and allies across the world, our support remains consistent with the core values and strong support for human rights expressed by the American people.[62]

For many years, NGOs have also been vocal in advocating for more restraint in U.S. arms transfers and for better methods of ensuring accountability. The Biden administration's enhanced focus on human rights implications provided a window in which long-standing advocacy for policy change has intensified. In 2021, a group of NGO experts representing various organizations published an article calling for significant change in U.S. arms-transfer policy, as evident in the following text from the article:

> End-use monitoring programs do not currently monitor whether U.S.-origin weapons are being used in the commission of human rights abuses, violations of IHL, or other acts of civilian harm. Biden should make clear in the new presidential decision directive that use of U.S.-

---

[61] U.S. Senate, 2024.

[62] Van Hollen, 2024.

origin defense articles or services in the commission of gross violations of human rights or IHL constitute end-use violations under the Arms Export Control Act.[63]

The authors argued that new EUM programs that could track the actual use of U.S.-origin items, including human rights violations, IHL violations, and civilian casualties, should be developed to address this gap.[64] Furthermore, in December 2023, 14 organizations focused on humanitarian causes, human rights, and the protection of civilians cosigned a letter to Secretary Austin that condemned the absence of robust U.S. monitoring of Israel's compliance with IHL and argued that the civilian harm resulting from U.S.-supported Israeli military operations was undermining commitments conveyed in the CHMR-AP and inconsistent with CAT Policy, as evident in the following text from the letter:

> Conditions ensuring compliance with IHL and human rights, including the facilitation of humanitarian assistance, should be the baseline of all U.S. assistance to any country. Continuous monitoring of possible IHL violations in Israeli operations is also critical for good-faith implementation of the CAT Policy and CHMR-AP, including to inform risk assessments and determinations regarding the "more likely than not" standard.[65]

There appear to be several reasons these bills have not yet become law. Generally, support for the application of additional oversight of partner use of U.S.-supplied arms is divided along partisan lines, and debates over U.S. arms-transfer policies have become increasingly polarized since the beginning of the war in Gaza. Concerns about retaining U.S. competitiveness against adversaries who do not apply a civilian harm lens to arms transfers

---

[63] Annie Shiel, Seth Binder, Jeff Abramson, William Hartung, Rachel Stohl, Diana Ohlbaum, Adam Isacson, Brittany Benowitz, and Daniel R. Mahanty, "Toward a More Responsible US Arms Trade Policy: Recommendations for the Biden-Harris Administration," *Just Security* blog, January 19, 2021.

[64] Shiel et al., 2021.

[65] Airwars, Amnesty International USA, Anera, et al., civilian harm in Gaza, letter to Lloyd J. Austin III, December 20, 2023.

are also prominent. According to one expert who has been closely tracking legislative initiatives related to arms transfers, some legislators have been encouraged by the CHIRG and regard this as being "partially responsive" to their concerns, and, therefore, there is a willingness to be patient and allow time for these new processes to yield results before pushing for additional legislative change.[66]

Conclusion

This chapter has explored how recent developments in U.S. policy positions regarding the responsibility to mitigate civilian harm have shaped the context in which EUM is being examined. Because U.S. policy has made more-expansive commitments to mitigate civilian harm and make human rights a more significant consideration in arms-transfer policy, the narrow scope of EUM and its exclusive focus on technological security appears increasingly incompatible with U.S. policy. Legislative proposals to expand the scope of EUM highlight congressional dissatisfaction with the status quo and reflect a desire for more visibility into partner usage of U.S.-supplied arms. Recent legislative proposals reflect a growing push to broaden the goals of EUM by including new requirements that would ensure compliance with international law and human rights standards, and these proposals make direct reference to broader DoD and U.S. policies on civilian harm mitigation and the importance of human rights considerations.

Through proposals to amend the AECA, which serves as the legal basis to define the scope and purpose of EUM, and through efforts introduce new reporting requirements to Congress on the adherence of recipient countries to IHL, these efforts collectively represent a growing desire for new legislative oversight mechanisms and more accountability. In the next chapter, we examine how a program that addresses these concerns could be implemented, discuss the new roles and variety of skills that would be required to undertake this new mission, and highlight some of the challenges that should be anticipated.

---

[66] Legislative expert on U.S. arms-transfer policies, interview with the authors, August 26, 2024.

CHAPTER 4

How an Expanded End-Use Monitoring Policy Might Include Operational Monitoring

To clearly demonstrate the implications of expanding the scope of EUM, our analysis considers the gaps that are exposed by a broadened mission. We have also identified some of the challenges that are likely to arise in the context of a scope expansion to EUM. In this chapter, we highlight some considerations in terms of the development of new skills for the execution of a more complex EUM mission and some of the policy, legal, and strategic challenges that we anticipate decisionmakers will need to confront. Considering these challenges can help inform the development of the workforce and the policies required to resource and oversee an EUM policy with an expanded scope.

Table 4.1 summarizes our findings regarding the major differences between the existing EUM undertaken as part of Golden Sentry and what we envision in a future EUM program that includes operational EUM and better reflects the new policy environment. The purpose of this table is to illustrate the differences between these efforts in terms of scope, operational environment, tasks, and methods. This comparison helps to demonstrate why the future DoD workforce that will undertake operational EUM will need skills that are different from those implementing Golden Sentry and why the organizational and managerial approaches that have been developed to implement Golden Sentry will not necessarily be appropriate or relevant for this new mission.

TABLE 4.1
Golden Sentry vs. Operational End-Use Monitoring

Category	Golden Sentry Program (Existing)	Operational EUM (Proposed)
Scope and focus	• Concerned with third-party transfer (diversion), physical security or storage, and technology security	• Concerned with operational employment, IHL, law of armed conflict, and civilian harm
Operational environment	• Mainly occurs in peacetime and permissive conditions	• Likely to occur in conflict settings and hostile environments
Tasks	• Review transfer agreements • Ensure compliance with specific physical security or accountability procedures	• Review, respond to, and investigate reports of civilian harm purportedly carried out with U.S.-origin defense articles • Assess whether weapons have been used in accordance with IHL and civilian harm mitigation best practices
Methods	• Conduct EEUM checks through planned or coordinated visits to the partner nation's installations • Verify in-country receipt of EEUM-designated defense articles by serial number	• Monitoring might require eyewitness interviews, collection of human intelligence, review of satellite imagery, forensic analysis, and verification through geolocation
Reporting requirements	• Report possible violations to DoS (PM and RSAT), DSCA, and the CCMD	• Report possible violations to DoS (PM and RSAT), DSCA, and the CCMD

NOTE: RSAT = Office of Regional Security and Arms Transfers.

New Skills for a New Mission

We interviewed NGO experts with significant expertise in the investigation of reports of civilian harm resulting from partner use of U.S.-origin defense articles. These conversations provided some insights into the process, skills, and experience that are required to undertake investigation into operational EUM. In describing the general standards of an investigation, one expert said, "We are trying to prove that a certain thing happened at a certain place at a certain time. What is the available evidence, and can we gather enough evidence where we can prove, above general journalistic standards, that something happened?"[1] To establish what one expert referred to as a "preponderance of evidence" for a human rights violation, the standard will include interviewing eyewitnesses and reviewing satellite imagery, but the expert explained that there is not necessarily a standard checklist that can be applied to every case.[2]

There is a consensus among experts that to document these incidents in a rigorous way is very "fact intensive."[3] Information, including victims' names, ages, and death certificates, are collected to demonstrate that something unlawful occurred. In some instances in which U.S.-origin defense articles have reportedly contributed directly to civilian harm, weapon fragments definitively identified as U.S.-origin were recovered from the scene.[4] However, experts explained that often this is not the case, and one NGO expert claimed,

> It is rare that you get a fragment with a serial number that you can trace back. . . . Largely speaking, the bombs are destroyed on impact. On EUM, it might be a mistake, if talking about human rights law, to say that you have to find definitive evidence that a certain bomb was responsible for a certain strike.[5]

According to our interviews with experts in investigations related to civilian harm, several skills, including satellite imagery analysis, use of geolocation technology, forensic analysis, postblast analysis associated with explosive ordnance, human intelligence collection, eyewitness interviews, and legal analysis, are required for investigations. The professionals who

---

[1] NGO expert on civilian harm investigations, interview with the authors, January 25, 2024.

[2] NGO expert on civilian harm investigations, interview with the authors, January 25, 2024.

[3] Forum on the Arms Trade, 2024.

[4] Amnesty International, "Israel/OPT: US-Made Munitions Killed 43 Civilians in Two Documented Israeli Air Strikes in Gaza—New Investigation," December 5, 2023b.

[5] NGO expert on civilian harm investigations, interview with the authors, January 25, 2024.

are engaged in this work tend to come from a wide variety of backgrounds, including those who worked for humanitarian organizations, have international law experience, or were former military and intelligence professionals. NGOs doing these investigations typically approach their investigations by pairing their in-house expertise with the recruitment of locals who have the requisite language skills, cultural knowledge, and social networks to facilitate investigations.[6]

In a 2021 *New York Times* article titled "The Human Toll of America's Air Wars," Azmat Khan, an investigative journalist, described her approach to investigating civilian harm incidents that the U.S. military had deemed to be "credible" through independent ground reporting via interviews with survivors.[7] The following account provides glimpses into the complexity of investigating civilian harm and the wide variety of skills associated with the complex task:

> Before visiting a credible site, I would analyze the document to identify central details about the allegation, the intelligence, what the military concluded the target was, how it was authorized to be bombed, what was observed and recorded in footage, chats and mission reports, the casualties assessed and other details. Next, I would research the coordinates the military provided. I would analyze that location in historical satellite imagery, both from before the date of the strike and after it, to identify potential impact areas and examine whether anything matched the target description in the document, or whether it was possible that the coordinates were incorrect.
>
> I decided to start with Mosul. I wanted to sample a large number of sites, and this would take time. Mosul was a place where I had developed the kind of reporting network that would enable me to work safely for several months. To prepare for my visit, I hired two students from the University of Mosul's department of translation, Momen

---

[6] NGO expert on civilian harm investigations, interview with the authors, January 25, 2024.

[7] Azmat Khan, "The Human Toll of America's Air Wars," *New York Times*, December 19, 2021.

Muhanned and Zainab Alfakheri, and trained them in some of the basic techniques of investigative journalism.[8]

Khan's account goes on to describe how she used crowdsourced mapping tools to understand the local area, open-source material about the incident, social media websites, and internet archiving sites for videos and photos, which were then cross-checked for matches with an existing repository of videos of bombings uploaded by the coalition.

Policy Challenges

To implement an expansion of EUM that includes monitoring the battlefield use of U.S.-origin defense articles and seeks to ensure compliance with international law and IHL, the U.S. government will need to develop new monitoring processes and investigative capacities that involve a variety of new skills (as discussed above) and leverage existing experience on security assistance and partner-nation relationships. Investigating harm linked to partners' use of U.S.-origin weapons creates several new challenges. Unlike harms that are directly caused by U.S. military operations, harms caused by partners will often occur in locations where the United States lacks significant military or diplomatic presence. In such cases, the U.S. government will need to develop approaches to gain access to information and networks that could corroborate claims of civilian harm. The volume of potential harms caused by a broad set of partners—as opposed to a narrower scope of incidents caused strictly by the U.S. military—is likely to create capacity challenges for investigations. The nature and purpose of pretransfer agreements, as a mechanism to codify partner commitments with respect to operational use, will need to be considered. Decisions about the appropriate evidentiary standards to be applied in new policies will significantly shape the scope of monitoring responsibilities that accompany this mission and determine the necessary resources. These policy challenges are discussed at greater length below.

---

[8] Khan, 2021.

Access

To implement a monitoring program that addresses operational usage and supports thorough investigations, DoD will need robust mechanisms to secure physical access to and visibility into the areas and operations in question to make a more thorough investigation possible. These mechanisms will likely involve the development of new provisions in arms-transfer agreements that explicitly address the responsibilities of partners to provide data and information that can inform U.S. assessments and to facilitate access for U.S. personnel to conduct investigations for situations in which compliance with U.S. or international law is in question. NGOs have developed methods of conducting investigations through a combination of remote direction and hiring on-the-ground teams. However, there are certain types of investigations related to operational EUM that NGOs simply cannot undertake that must become part of a U.S. government program if they are to be undertaken at all. An NGO expert explained,

> There are investigations we can't do and that are unsafe to us. To talk about what type of investigations [the U.S. government] can do—U.S.-provided cluster munitions to Ukraine, who promised to use them in rural areas with few civilians. If that was the deal and someone wanted to do "OEUM" [operational EUM] with that, we could not investigate that. We would have to wait until after the border changes to figure out what we could do. [The U.S. government] could put someone in the HQ [headquarters] of the units firing DPICM [dual-purpose improved conventional munition] at Russian lines. They would not have forensic view but could have some oversight in an operational environment.[9]

In DoS's report on NSM-20 delivered to Congress in May 2024, there were many indications that the lack of physical access impeded U.S. efforts to fully evaluate Israel's compliance with the requirements of the national security memorandum. The report explicitly stated that "Israel has not shared complete information to verify whether U.S. defense articles covered under NSM-20 were specifically used in actions that have been alleged

---

[9] NGO expert in civilian harm investigations, interview with the authors, January 25, 2024.

as violations of IHL or IHRL [international human rights law] in Gaza, or in the West Bank and East Jerusalem during the period of the report," which underscores the U.S. government's significant reliance on partner self-reporting. The information that Israel shared regarding the use of U.S.-supplied weapons was characterized as "incomplete": "Israel has, upon request, shared some information on specific incidents implicating IHL, some details of its targeting choices, and some battle damage assessments."[10] The report also confirmed that the lack of U.S. personnel on the ground made it "difficult to assess or reach conclusive findings on individual incidents" and concluded that even though Israel had undertaken some steps to implement IHL obligations, the U.S. government "lacks full visibility into Israel's application of these principles and procedures."[11]

The role of the United States as a supplier of weapons to a country involved in active conflict might also complicate the U.S. government's ability to conduct objective investigations that incorporate a variety of sources and on-the-ground reporting and might interfere with its ability to engage with individuals reporting on harms caused by the U.S. government. These people might not be willing to speak directly to the U.S. government about incidents, out of fear of retribution. As one experienced NGO expert explained,

> People are more likely to talk to us than the [the U.S. government]. It is safe to talk to us because we have been doing this for 60 years. . . . [The challenge of] making people feel safe and comfortable if they talked to [the U.S. government] would be something to overcome.[12]

From speaking with NGOs about the processes that they use to investigate these claims, we realized that both U.S. officials and NGOs have unique advantages and disadvantages in this process. U.S. officials will be privy to data and intelligence that NGOs will not have access to, whereas NGOs

---

[10] DoS, *Report to Congress Under Section 2 of the National Security Memorandum on Safeguards and Accountability with Respect to Transferred Defense Articles and Defense Services (NSM-20)*, May 10, 2024.

[11] DoS, 2024.

[12] NGO expert, interview with the authors, January 31, 2024.

will often have extensive human networks, language skills, cultural familiarity, and trust established with the local population that U.S. personnel will not be able to immediately able to replicate. Gaining the full operational perspective and collecting necessary forensic data will likely require cooperation between both U.S. personnel and NGOs, with both leveraging their respective strengths and access.

Pretransfer Agreements

Existing pretransfer agreements require the recipient to acknowledge willingness and ability to apply international law, but to institutionalize an expanded program, new language would need to be integrated into pretransfer assurances.[13] The need for pretransfer assurances related to civilian harm mitigation is addressed in Action 9.i. of the CHMR-AP with reference to "pre-transfer assurances that document willingness to apply appropriate civilian harm mitigation measures," but the content of the pretransfer assurances will need to be considered carefully.[14] Although some monitoring can likely be undertaken remotely, agreements will likely need to include a provision that allows access to U.S. investigators for instances in which reports of civilian harm have been deemed credible and require further investigation. In the absence of physical access, investigations might need to rely heavily on external reporting from NGOs, journalists, and partner-supplied self-assessments, which are likely to be deemed inadequate for a definitive finding.

Researchers have highlighted the opportunity to better leverage pretransfer agreements in shaping partner behavior with respect to weapon use. In a joint report published by the Center for Civilians in Conflict and the Stimson Center in 2018, the authors wrote,

> Strengthening terms of sale and end-use monitoring: Maintaining basic access, oversight, and visibility into the use of US sold defense items should be a part of the weapons sales lifecycle. The US government agencies involved in arms sales under-utilize contractual agree-

---

[13] U.S. official, interview with the authors, February 9, 2024.

[14] DoD, 2022b.

ments as a key instrument for controlling and reviewing the use of defense items.[15]

Per Section 2.2.f. of DoDI 3000.17, DSCA is assigned the responsibility of incorporating into defense article and service-transfer agreements

> an understanding that recipients of security cooperation and security assistance will provide pre-transfer assurances that document ally and partner willingness and ability to apply effective civilian harm mitigation practices, including when employing U.S.-origin defense articles and services.[16]

If EUM policy becomes broadened to include monitoring of operational use through a statutory requirement, pretransfer assurances might need to be adapted to reflect the relevant requirements for its implementation.

Evidentiary Standards

Another policy challenge that will need to be addressed is the question of evidentiary standards. Decisions about evidentiary standards will directly shape the monitoring responsibilities that accompany this mission and the necessary resources. In this context, *evidentiary standards* refers to the threshold that will be applied in deciding what facts are true and what kinds of evidence will be required and admitted to inform those judgments. This is also sometimes referred to as a *standard of proof*.[17] Although the concept of evidentiary standards derives from a legal context, we classify this as a policy challenge because the application of these standards will inform arms-transfer policy decisions as opposed to serving as the basis to build a legal case against a foreign country. However, the principle of determining what level of certainty and evidence will be required to prove facts or be

---

[15] Daniel Mahanty and Annie Shiel, *With Great Power: Modifying US Arms Sales to Reduce Civilian Harm*, Center for Civilians in Conflict and Stimson Center, 2018, p. 9.

[16] DoD, 2023b, p. 11.

[17] *Standard of proof* refers to a degree or a range of certainty within which facts need to be established. See Aniruddha Rajput, "Standard of Proof," in Anne Peters, ed., *Max Planck Encyclopedias of International Law*, Oxford University Press, February 2021.

considered sufficient is directly relevant to the implementation of operational EUM.

DoDI 3000.17 specifies that the evidentiary standard when assessing reports that U.S. military operations have resulted in civilian harm will be the following "more likely than not" standard:

> The assessment's results will indicate whether, based upon the information available at the time of the assessment and depending on the scope of the assessment, civilian casualties or damage to or destruction of civilian objects more likely than not resulted from U.S. military operations.[18]

The DoDI also references the complexities of gathering information during hostilities, explaining that "the 'more likely than not' standard also reflects the reality that information during military operations, including with respect to the outcomes of operations, is often lacking or incomplete."[19] The absence of complete information is likely to affect the investigations associated with operational EUM and might pose even greater challenges than investigations of U.S. military operations. The inherent challenges involved in collecting information on partner operations must be weighed against the benefits of establishing a consistent standard.

Interagency Coordination

DoS and DoS are both developing guidance for responding to allegations that partners have committed civilian harm. However, the CHIRG and pending interim DoD guidance are being designed as two separate processes. Although DoS and DoD do meet on a recurring basis to discuss responses to reports of civilian harm by ally or partner forces, there is no publicly available guidance that outlines coordination requirements between the

---

[18] See DoD, 2023b, p. 29. The "more likely than not" standard is also the same standard adopted in the CAT Policy to guide determinations regarding arms transfers (i.e., if it appears more likely than not that arms will be used by the recipient to commit serious violations of international humanitarian or human rights law, the transfer is prohibited).

[19] DoD, 2023b, p. 29.

CHIRG process and the pending DoD process or that clarifies each department's respective contributions.[20] The fact that both departments have distinct authorities for the transfer of defense articles makes establishing an interagency approach somewhat complicated, but U.S. officials working at the intersection of civilian harm mitigation and U.S. arms-transfer policies have acknowledged that if parallel processes emerge without sufficient alignment, it is likely to cause confusion in the field and might complicate the goal of collecting and investigating all credible reports in a timely manner. As one U.S. official said, "There may be cases where it's difficult to determine whether Title 10 or Title 22 authorities were used to transfer the articles in question."[21]

Legal Challenges

We anticipate several legal challenges in implementing a policy of operational EUM. NSM-20, as a policy, creates new requirements that, in turn, will raise a number of legal considerations. NSM-20 requires partners to provide credible and reliable written assurances from a representative of the recipient country that the recipient country will "use any such defense articles in accordance with international humanitarian law and, as applicable, other international law."[22] Following these assurances, the Secretaries of State and Defense are then required to assess whether usage has been found to be inconsistent with "established best practices for mitigating civilian harm, including practices that have been adopted by the United States military, and including measures implemented in response to the Department of Defense's Civilian Harm Mitigation and Response Action Plan."[23] This section will explore challenges inherent in requiring IHL compliance as a condition in arms transfers—in particular, the complexity of assessing whether allies and partners have appropriately balanced the legal principles

---

[20] For information on periodic DoD-DoS meetings, see Blair, 2022.

[21] U.S. official, interview with the authors, February 9, 2024.

[22] White House, 2024.

[23] White House, 2024.

of humanity and military necessity. We then consider how other regimes governing weapon use, including those enshrined in the DoD *Law of War Manual* will shape legal considerations impacting operational EUM.

International Humanitarian Law Principles

When considering the challenge of promoting partner compliance with IHL as a requirement in arms transfers, it is important to recognize the dynamic nature of international law and the reality that, as a legal regime, it is a reflection of shared norms and results from ongoing calculations about national security and state interest. In other words, as international law scholar Michael Schmitt put it: "International law thus reflects the goals of the states consenting to be bound by it."[24] Inherent in these calculations is an effort to balance norms that are in tension with each other, such as those of *humanity* and *military necessity*. Schmitt also addresses this issue of balance, underscoring the importance of contextual factors, in the following:

> Of course, all policy decisions are contextual in the sense of being based on past, existing, or anticipated circumstances. When circumstances change, the perceived sufficiency of a particular balancing of military necessity and humanity may come into question. In response, states reject, revise, or supplement current IHL or craft new norms.[25]

Assessing whether the principle of *military necessity* has been appropriately applied by partners is also likely to be fraught, because this principle is based on a perception regarding the nature of the threat and what is considered absolutely necessary to confront it. *Military necessity* refers to the requirement for the application of armed force (in accordance with the other rules of the law of armed conflict) to achieve legitimate military objectives.[26] States often accept high levels of civilian casualties from situa-

---

[24] Michael N. Schmitt, "Military Necessity and Humanity in International Humanitarian Law: Preserving the Delicate Balance," *Virginia Journal of International Law*, Vol. 50, No. 4, May 2010, pp. 798–799.

[25] Schmitt, 2010, p. 799.

[26] See Schmitt, 2010. When crafting IHL, states insist that legal norms not unduly restrict their freedom of action on the battlefield, such that national interests might be

tions in which they regard the threat to be existential. This threat perception shapes the context for the application of the principle of *proportionality*: The legality of an attack is evaluated according to what the commander knew about the expected harm in relation to the military advantage anticipated.[27] Determining adherence to these principles on behalf of partners and concluding whether their own balancing of these principles is sufficient to reflect the best practices of civilian harm mitigation would require extensive knowledge of their operational context.

Applicable Legal Regimes Governing Weapons Use

The best practices adopted by the U.S. military should include relevant aspects of the DoD *Law of War Manual* that directly address what is permissible with respect to the use of inherently indiscriminate weapons and the rules governing use of incendiary weapons against military objectives that are located within a concentration of civilians.[28] The use of certain types of weapons will make operational EUM especially challenging because of the complexity of the legal regimes surrounding the weapons' use. For example, an *incendiary weapon* is defined as "any weapon or munition that is primarily designed to set fire to objects or to cause burn injury to persons through the action of flame, heat, or combination thereof, produced by a chemical reaction of a substance on the target."[29] These weapons are generally regulated under the Protocol on Prohibitions or Restrictions on the Use of Incendiary Weapons (Protocol III) of the Convention on Certain Conventional Weapons.[30] However only "pure" incendiaries, such as napalm or

---

affected. The principle of military necessity constitutes the IHL mechanism for safeguarding this purpose.

[27] Beth Van Schaack, "Evaluating Proportionality and Long-Term Civilian Harm Under the Laws of War," *Just Security* blog, August 29, 2016.

[28] See DoD, *Law of War Manual*, updated July 2023a; and Protocol on Prohibitions or Restrictions on the Use of Incendiary Weapons (Protocol III), adopted at Geneva, Switzerland, October 10, 1980. According to Article 2, Section 1 of Protocol III, "It is prohibited in all circumstances to make the civilian population as such, individual civilians or civilian objects the object of attack by incendiary weapons."

[29] Protocol on Prohibitions or Restrictions on the Use of Incendiary Weapons, 1980.

[30] Protocol on Prohibitions or Restrictions on the Use of Incendiary Weapons, 1980.

the type of incendiary bombs used in World War II and Korea, are regulated as incendiary weapons under international law, and munitions that might have "incidental incendiary effects" are legally excluded from the definition.

One example of a weapon that falls outside Protocol III but has been documented to cause significant civilian harm is white phosphorous.[31] White phosphorous is intended primarily for illuminating a target or masking friendly force movement by creating a smoke screen and is considered a multipurpose munition. In October 2023, Amnesty International reported on Israel's alleged unlawful use of white phosphorus in southern Lebanon.[32] DSCA's notes, addressing the terms regarding appropriate use for inclusion in a letter of offer and acceptance, specify that "white phosphorus munitions are to be used only for purposes such as signaling and smoke screening."[33] Operational EUM of this particular weapon to confirm compliance would require insight into the recipient's intent and various other contextual factors to determine whether the incidental incendiary effects were being exploited unlawfully. Other analysts considering the same case of Israel's use of the weapons in Lebanon concluded that, "as always, the application of the law of armed conflict is contextual. Whether the use of white phosphorous violates the law depends on the circumstances and manner of its use, which, in the present case, remain unclear."[34]

Strategic Challenges

The strategic challenges associated with monitoring the operational end use of weapons are manifold. Arms transfers have long been seen as crucial U.S. foreign policy tools that have significant implications for regional and global security. In FY 2023 the total value of transferred defense articles

---

[31] World Health Organization, "White Phosphorus," fact sheet, January 15, 2024.

[32] Amnesty International, "Lebanon: Evidence of Israel's Unlawful Use of White Phosphorus in Southern Lebanon as Cross-Border Hostilities Escalate," October 31, 2023a.

[33] See Appendix 6 of the SAMM (DSCA, "LOA Notes Listing," in *Security Assistance Management Manual*, 2024b).

[34] Kevin S. Coble and John C. Tramazzo, "Israel-Hamas 2023 Symposium—White Phosphorus and International Law," *Articles of War* blog, October 25, 2023.

and services and security cooperation activities conducted under the FMS system was $80.9 billion.[35] The scale of U.S. arms transfers is enormous, and the strategic and economic implications of policies that impose new restrictions shaping the U.S. role in the global arms supply are enormous. Even though changes in CAT Policy during previous administrations have included vows to exercise "restraint against the transfer of arms that would enhance the military capabilities of hostile states, serve to facilitate human rights abuses or violations of international humanitarian law, or otherwise undermine international security," it has become clear that pronounced changes in policy with respect to long-term U.S.-origin arms purchasers have been very rare, even when the values of the United States and its partners appear to diverge.[36]

Historically, intersections in security interests with partners and considerations regarding local balances of power have proven to be enduring drivers of arms-transfer decisions, and these factors appear to contribute to continuity over time, even as political factors and policy rhetoric evolve.[37] Furthermore, conditionality in arms transfers is often considered a risk in the context of strategic competition. Some argue that the more conditions applied by the United States in its arms transfers, the more attractive the "no-strings" supplier might become. However, other research indicates that states consider a variety of factors when considering partners from which to purchase arms.[38] Conditionality may be part of the equation, but so are considerations on diversification, dependence, and interoperability of

---

[35] DoD, "Fiscal Year 2023 U.S. Arms Transfers and Defense Trade," fact sheet, January 29, 2024.

[36] White House, 2023.

[37] A. Trevor Thrall, Jordan Cohen, and Caroline Dorminey, "Power, Profit, or Prudence? US Arms Sales Since 9/11," *Strategic Studies Quarterly*, Vol. 14, No. 2, Summer 2020; Melissa Shostak and Cortney Weinbaum, "How China Is Building Influence Through Arms Sales," *National Interest*, December 7, 2022; Çağlar Kurç, "No Strings Attached: Understanding Turkey's Arms Exports to Africa," *Journal of Balkan and Near Eastern Studies*, Vol. 26, No. 3, 2024.

[38] Elizabeth Dent and Grant Rumley, "How the U.S. Used Arms Sales to Shift Saudi Behavior," *Defense One*, September 4, 2024.

systems.[39] Furthermore, research suggests that the influence relationship between arms exporters and recipients is complex and that exporting states may not aways derive the influence from arms transfers that they seek.[40]

An expanded scope for EUM to include operational monitoring would mark a departure from the standard practice of EUM programming to date. An EUM program that monitors partners operations will provide the United States with a greater understanding of how partners employ U.S.-origin weapons in conflict. The strategic implications of conditioning arms transfers on civilian harm considerations remain a contentious topic, however. When U.S. officials consider options for responding to reports of civilian harm, they will have to consider both the strategic risks resulting from civilian harm that results from partner use of U.S.-origin weapons and those that might result from the application of more conditionality in arms-transfer relationships.

Implementation Roles for the Department of Defense

DoD officials have expressed concerns that a program explicitly designed to carry out operational EUM exceeds both the existing resources and the established roles of SCOs.[41] Although DoD policy acknowledges that SCOs have a responsibility to report the use of U.S.-origin defense articles transferred through FMS against anything other than legitimate military targets, comprehensive evaluation of partners' operations to identify noncompliance will likely represent a significant expansion to the routine tasks associated with Golden Sentry EUM. DSCA has suggested that reporting these

---

[39] Elias Yousif, *"If We Don't Sell It, Someone Else Will": Dependence and Influence in U.S. Arms Transfers*, Stimson Center, March 2023; C-SPAN, "Arms Control Association Discussion on U.S. Arms Transfers," video, June 7, 2024.

[40] Yousif, 2023. Also, see T. V. Paul, "Influence Through Arms Transfers: Lessons from the U.S.-Pakistani Relationship," *Asian Survey*, Vol. 32, No. 12, December 1992; and Keith Krause, "Military Statecraft: Power and Influence in Soviet and American Arms Transfer Relationships," *International Studies Quarterly*, Vol. 35, No. 3, September 1991.

[41] DoD officials, interviews with the authors, October 25, 2023, and January 16, 2024.

incidents might be aligned to role that is better located elsewhere, given the limited resources assigned to EUM and the limited visibility that SCOs' duties provide them into foreign partners' use of equipment.[42] These are important considerations that must be reflected in program design and the assignment of new roles and responsibilities. According to established DoD policy and our analysis of the requirements of this new mission, there are elements that could be performed by SCOs and elements that should reside elsewhere. To establish this capability, new roles in terms of guidance, oversight, direct implementation, and integration will need to be assumed by a variety of DoD components. We offer some initial thoughts on possible roles below, according to existing policy guidance and established areas of expertise.

Roles for the Office of the Secretary of Defense

The Office of the Secretary of Defense (OSD) has policy guidance responsibilities that can shape the design and implementation of a new operational EUM program. Specifically, the Deputy Assistant Secretary of Defense for Global Partnerships has oversight of security cooperation policy (including policies regarding the transfer of defense articles), relationships with DoS in interagency matters related to security sector assistance, and leadership in implementing civilian harm mitigation efforts related to allies and partners. The Deputy Assistant Secretary of Defense for Irregular Warfare and Counterterrorism, Special Operations and Low-Intensity Conflict, is the policy lead for broader issues associated with civilian harm mitigation and response and the implementation of the CHMR-AP. Together, these offices could provide guidance on the development and implementation of a new operational EUM program that integrates civilian harm mitigation objectives and identifies necessary contributions from the developing DoD civilian harm mitigation community. The initial steps we envision for OSD in establishing the first phase of this program are discussed at greater length in Chapter 5.

---

[42] See, Blair, 2022, p. 31.

Roles for the Defense Security Cooperation Agency

DSCA has been assigned the role of supporting the implementation of Civilian Harm Mitigation and Response Baselines of Allies and Partners (CBAPs) and ensuring that processes for DoD security cooperation programs are in place to assess, monitor, and evaluate the ability, willingness, norms, and practices of allies and partners to implement appropriate and effective civilian harm mitigation and reduction practices.[43] Given this responsibility, and the expanded role DSCA will assume in FY 2025 for training and managing SCOs through the newly established Defense Security Cooperation Service,[44] DSCA could help to strengthen relationships between the security cooperation and across the security sector assistance communities, including those involved in EUM and those engaged civilian harm mitigation, to cultivate a new subset of expertise in DoD that integrates security sector assistance knowledge and civilian harm mitigation best practices and that is developed specifically for the implementation of monitoring partner operations that use U.S.-origin weapons. The Civilian Protection Center of Excellence (CP CoE) could offer valuable expertise in this regard through convening the civilian harm mitigation community of practice and through documenting and disseminating lessons from the broader community.

Roles for Security Cooperation Organizations

The existing resources, staffing, and tasks assigned to SCOs are not sufficient to facilitate visibility into partner operations. However, SCOs' roles and access to partners' defense organizations might make them well suited to implement an expanded EUM program, conditional on additional resources and the development of training tailored to these new responsibilities.[45] The role of the SCO as a facilitator of access is particularly signifi-

---

[43] See Section 2.2.D of DoD, 2023b. CBAPs are a requirement introduced in Objective 9 of the CHMRAP to support the integration of CHMR into DoD security cooperation programs

[44] U.S. Code, Title 10, Chapter 16, Section 384, Department of Defense Security Cooperation Workforce Development.

[45] The Certification 2.0 Program, an element of the DoD Security Cooperation Workforce Development Program that is prescribed by U.S. Code, Title 10, Section 384, is

cant. As laid out in the Defense Security Cooperation University publication *Security Cooperation Management,* the SCO's relationship with the partner nation represents a unique resource:

> The SCO needs access to host nation counterparts to ensure USG [U.S. government] objectives can be met. Central to this access are the SCO's relationships with the host nation. In some cases, personal relationships are required before professional interactions are possible. Developing these crucial relationships takes time and patience—the security cooperation business is not an overnight enterprise. The SCO's fundamental task is to effect USG foreign policy and, in many cases, to build host nation capabilities and capacities to meet future USG and host nation challenges.[46]

The SCO manages DoD security cooperation programs under the guidance of the CCMD. To the extent that operational EUM might be considered part of the management of security assistance or security cooperation programs, it would make sense for SCOs to play a central role in this process. The SCO also plays a crucial role in synchronizing resources, programs, and policy priorities across the interagency, and this function would likely be important in the implementation of monitoring partners operational use of U.S.-origin weapons. The synchronizing role would likely include close cooperation with other personnel in the embassy who have a role in human rights oversight and especially with those who are assigned to contribute to the CHIRG progress.[47]

---

required to ensure that DoD personnel assigned to statutorily defined Security Cooperation Workforce (SCW) positions have the SCW competency-based training and experience necessary to carry out assigned security cooperation responsibilities. SCO training is being revised and expanded. If operational EUM were to be assigned to DoD, the Certification 2.0 program offers an opportunity to incorporate broader training relevant to the mission.

[46] Defense Security Cooperation University, "Security Cooperation Organizations Overseas," in *Security Cooperation Management,* 43rd ed., 2023b, p. 4-1.

[47] These relationships might span U.S. defense contractors and special DoD agencies; members of the U.S. embassy country teams associated with partner nations; a variety of DoS elements; the CCMD J5 country desk officer, other headquarters, and component command staff; members of the DSCA staff; representatives of the DoD imple-

However, the SCO role in operational EUM will look very different and might be less hands-on than the SCO role in Golden Sentry EUM. It is likely a synchronizing and integrative function, given SCOs' ability and access to not only liaise directly with in-country stakeholders but also reach back to other resources and tap into specialized expertise in the CCMD. Some tasks associated with this mission might be regarded by partners as intrusive at times and have the potential to strain bilateral relations, which might create challenges for SCOs and introduce tensions with other established SCO responsibilities. This is a matter that must be managed through policy direction and carefully developed training.

Roles for the Defense Intelligence Enterprise

Our interviews have highlighted the fact that ensuring partner compliance with U.S. policy and law in the operations will require a monitoring program that is explicitly designed to gain visibility into partner operations. According to DoDI 3000.17, the Office of the Under Secretary of Defense for Intelligence & Security must designate

> a responsible producer for analysis of civilians and civilian objects within the Defense Intelligence Enterprise through the Defense Intelligence Analysis Program to . . . [i]ncorporate foreign force [civilian harm mitigation capability] analysis into relevant product lines and mission sets, and support DoD Components and other [U.S. government] departments and agencies, as appropriate, by responding to requests for information.[48]

With additional guidance from OSD and DSCA regarding the specific intelligence requirements to implement this program, the intelligence produced to satisfy existing requirements in DoDI 3000.17 could be leveraged to gain insight into partner use and compliance with policy and legal requirements.

---

menting agencies; and senior DoD officials and agencies, such as OSD and the Joint Staff. See Defense Security Cooperation University, 2023b.

[48] DoD, 2023b, p. 13.

Roles for the Combatant Commands

There are clearly functions and roles associated with this new mission that will be best matched with expertise and established responsibilities resident in CCMDs. The roles being established in CCMDs to implement DoDI 3000.17 are relevant, particularly those of civilian harm assessment cells (CHACs). CHACs will include personnel with expertise in intelligence; joint fires; civil-military relations; and poststrike assessment and analysis, in addition to those with an understanding of the language, region, and culture relevant to the area of operations. CHACs will also have access to legal advice from command counsel. Operational EUM will require involvement of personnel with expertise in various areas, including targeting processes, weaponeering, and collateral damage estimation,[49] and the legal and doctrinal expertise of strike cell judge advocates general (JAGs), who play a role in ensuring U.S. military compliance with the law of armed conflict and the rules of engagement during targeting.[50] This suggests that a role for CHACs to support this new mission would be well aligned to the skills of constituent personnel. CHACs might be well suited to address many of the gaps that would result from a program primarily implemented by SCOs. However, the existing human capital and resource allocation for CHACs has been made primarily for the integration of civilian harm mitigation into U.S. military operations, so personnel cannot be simply reallocated toward a focus on partner operations or assigned expanded responsibilities without risking a failure to meet CHMR-AP objectives. Through the provision of additional personnel, the role of the CHAC could be expanded over time to contribute to a mission focused on monitoring partner operations.

---

[49] Joint Chiefs of Staff, *No-Strike and the Collateral Damage Estimation Methodology*, Chairman of the Joint Chiefs of Staff Instruction 3160.01A, October 12, 2012.

[50] See Todd Huntley, "Airstrikes, Civilian Casualties, and the Role of JAGs in the Targeting Process," Lawfare Institute and Brookings Institution, August 22, 2022. JAG responsibility for establishing positive identification of targets as "lawful military objectives" could be called on to contribute to assessments of partner compliance with U.S. policy requirements and international law.

Conclusion

Instituting measures to conduct monitoring of partners' operational use of U.S.-supplied weapons in conflict zones involves significant challenges. Some policy challenges will center on ensuring access to areas and obtaining reliable information on partner operations, both through DoD's independent collection abilities and through leveraging data collected by partners and NGOs. The pretransfer agreements that exist to govern arms transfers might need to be revised to include additional provisions for access and data sharing. Establishing the appropriate standards for evidence and proof in the context of this program will require a high level of coordination between DoS and DoD. Furthermore, the strategic challenges involved in introducing new conditions into transfers agreements are substantial. In the context of strategic competition, increasing the conditions involved in arms transfers will inevitably create tensions at times between U.S. strategic and foreign policy goals and commitments to mitigate civilian harm. The implementation of this mission will require a broader and, in some cases, more technical skill set than Golden Sentry requires. Therefore, this is not a mission that is well suited to be exclusively carried out by SCOs. To be implemented effectively, this mission will require the sharing of responsibilities and close coordination between various components across DoD and DoS.

CHAPTER 5

Recommendations

In this chapter, we provide our recommendations to help DoD establish an operational EUM program. These recommendations have been developed to address the gaps and challenges identified in previous chapters. Because the expansion of EUM to address partners' operational end use of U.S. defense articles will involve the development of a new mission and new expertise, we recommend a phased approach to program expansion. This phased approach will allow DoD to collect important data that will help to define and shape future manpower and resource requirements.

Ultimately, these recommendations suggest ways in which DoD can gain greater insight into how partners employ U.S.-origin equipment—insights that can be used to inform longer-term policy considerations about which partnerships further U.S. strategic interests abroad. Given the scope of our study, our recommendations focus on steps that would improve DoD's own organic capacity to monitor and investigate civilian harm stemming from partners' use of U.S.-origin equipment. However, we recognize that other interagency stakeholders also play an important role in the consequential policy processes concerning arms-transfer policy. Therefore, we also include some discussion about how DoD should integrate its operational EUM program with related interagency processes, such as DoS's CHIRG process. After the U.S. interagency has better access to information on partners' use of U.S.-origin equipment, this information can shape decisions on how to engage with partners to prevent civilian harm and how to apply tailored conditionality to further U.S. strategic objectives.

Establish New Policy Guidance That Clarifies Terminology and Assigns Responsibilities for the Department of Defense's Monitoring of Partners' Employment of U.S.-Origin Defense Articles

DoDI 4140.66 refers to Golden Sentry as the sole DoD program supporting EUM, which has created misperceptions among policymakers and legislators about the program's focus.[1] The Golden Sentry program is resourced and structured to verify the physical security of U.S.-origin defense articles, not to monitor their operational use and investigate potential associated civilian harm. The term *end use* will remain misleading with respect to Golden Sentry's purpose unless clarified by explicit policy documentation. Therefore, we recommend that DoD adopt new policy guidance that defines an umbrella EUM process and clarifies the subobjectives of programs that fall under it. Specifically, a new DoDI should clarify that *EUM* refers broadly to the routine monitoring of partners' storage and their operational use of U.S.-origin defense articles to (1) ensure their physical security and (2) monitor and investigate civilian harm stemming from their use. This language will help clarify the scope of EUM and delineate between Golden Sentry and a distinct but related program focused on responsible operational end use, hereafter referred to as the proposed Arms Transfer Accountability Program (ATAP).

Figure 5.1 illustrates our recommended concept for a new DoDI that defines EUM as an umbrella term encompassing two distinct programs that both seek to mitigate the risks involved in U.S. arms transfers: Golden Sentry and the proposed ATAP. The new DoDI should clarify that Golden Sentry focuses on preventing unauthorized transfers of U.S.-origin equipment to third parties by ensuring safe storage, whereas ATAP focuses on monitoring usage among partners to ensure compliance with applicable legal regimes and pretransfer assurances and, when necessary, the investigation of reports of harm caused by partners' and allies' use of U.S.-origin equipment.

---

[1] DoD, *Registration and End-Use Monitoring of Defense Articles and/or Defense Services*, DoDI 4140.66, incorporating change 1, May 24, 2017.

FIGURE 5.1
End-Use Monitoring Umbrella Concept

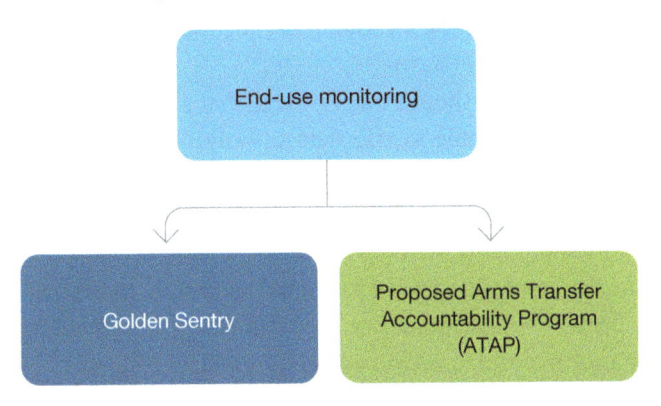

```
                    End-use monitoring
                    /              \
            Golden Sentry    Proposed Arms Transfer
                             Accountability Program
                                    (ATAP)
```

DoD should coordinate the design and implementation of the proposed ATAP with the interagency to ensure that it is integrated with other related efforts, such as the DoS CHIRG process. A new DoDI should assign clear responsibilities to DoD components with respect to information collection (e.g., responsibilities for investigations, intelligence collection, site visits) and describe how data collected through the proposed ATAP will support related interagency efforts to understand partner use of U.S.-origin weapons. Through consultation with DoS, DoD should also specify the extent to which DoD will endeavor to be proactive in this mission. A key question to be addressed is: Will DoD actively seek out information to confirm that operational usage adheres to legal and policy requirements? Or will DoD respond only to external reports that violations have occurred? The extent to which DoD actively monitors usage will be significant in determining human capital and other resources required for the program.

The new DoDI should also formalize the relationship between the EUM and civilian harm mitigation communities. Significant expertise regarding the investigation of allegations of civilian harm resides in DoD. Although investigations of partner use will be different, and, in many ways, more challenging than investigations of civilian harm stemming from U.S. military operations, DoD should seek to leverage this existing capability to the greatest extent possible. To formalize the relationship, DoD could convene a working group for the community of interest to evaluate and refine the implementation of the proposed ATAP. DoD could also consult its civilian harm mitigation experts—including relevant personnel from CHACs and subject-matter experts in the CP CoE—in the process of reviewing and triaging incoming reports of harm.

Finally, a new DoDI should clarify the global scope of EUM programs. DoDI 4140.66 is applicable to transfers to only the governments of Iraq, Afghanistan, and Pakistan. A new DoDI should adopt a global scope to align with the 2023 CAT Policy, which acknowledges that "worldwide" respect for human rights strengthens U.S. national security.[2] These doctrinal changes—alongside the incremental approach to scope expansion described below—will demonstrate a commitment to upholding the Foreign Assistance Act and to addressing congressional concerns.[3] New policy direction to shape the program should reflect likely requirements for effective legislative oversight. The reporting practices used to carry out Leahy vetting might prove to be useful as a model when developing this guidance.[4] Previous RAND research noted that improvement to the implementation of Leahy could be carried out through clarifying the scope of Leahy vetting requirements by updating DoD's formal guidance and supplementing it with informal guidance that addresses real-world questions.[5] There could be a role for these two types of guidance for the implementation of a new EUM effort to address operational use, given the new mission and the need to distinguish Golden Sentry scope and purposes from those associated with the proposed ATAP. The process of formalizing the purpose of

---

[2] White House, 2023.

[3] U.S. Code, Title 22, Chapter 31, Section 2304, Human Rights and Security Assistance, prohibits security assistance—which includes the sale and loan of defense articles—for any government that "engages in a consistent pattern of gross violations of internationally recognized human rights." For a more complete list of human rights–related provisions in the Foreign Assistance Act, see Paul K. Kerr and Michael A. Weber, *U.S. Arms Sales and Human Rights: Legislative Basis and Frequently Asked Questions*, Congressional Research Service, IF11197, February 29, 2024.

[4] See DoS, "About the Leahy Law," fact sheet, Bureau of Democracy, Human Rights, and Labor, January 20, 2021a.

[5] Michael J. McNerney, Jonah Blank, Becca Wasser, Jeremy Boback, and Alexander Stephenson, *Improving Implementation of the Department of Defense Leahy Law*, RAND Corporation, RR-1737-OSD, 2017.

Recommendations

this new program, including clearly defining all missions, tasks, and functions assigned, will serve as a basis for determining and validating manpower requirements—a key step in ensuring that the program is resourced effectively.[6]

Adopt a Phased Approach to Resourcing a Program Scope Expansion

The adoption of a new DoDI will require OSD to shape the program scope and acquire additional resources to implement the proposed ATAP alongside Golden Sentry. To support a newly scoped EUM process that is consistent with DoD policy and processes for determining joint manpower requirements, we recommend that OSD take a phased approach to evaluating, requesting, and integrating new resources and personnel. A phased approach will permit OSD to develop a greater understanding of the skills and capabilities required to implement the proposed ATAP, develop validated manpower requirements, and determine and request the appropriate resources that accurately reflect the workload included in this new mission. Figure 5.2 illustrates our proposed phased approach to understanding and expanding resourcing and staffing EUM programs.

In Phase I, OSD should focus on data collection to inform the development of a clearer picture of the workforce skills and requirements associated with our proposed scope for the EUM process. Although DSCA has many years of experience implementing Golden Sentry, ATAP will require a unique set of skills, capabilities, and technologies to be implemented well.

To gather information on potential future requirements, we recommend that OSD designates staff in DSCA, the geographic CCMDs, and the CP CoE to begin implementing the new DoDI—inclusive of Golden Sentry and the proposed ATAP—with a temporarily limited scope during Phase I. For example, OSD could choose one region for initial implementation as a pilot. OSD should engage with the designated staff throughout the year to under-

---

[6] See Joint Chiefs of Staff, "Enclosure C—Joint Manpower and Program: Requirements Determination and Validation," in *Joint Manpower and Personnel Program*, Chairman of the Joint Chiefs of Staff Instruction 1001.01C 21, February 21, 2024.

FIGURE 5.2

Phased Approach to Resourcing a Program Scope Expansion

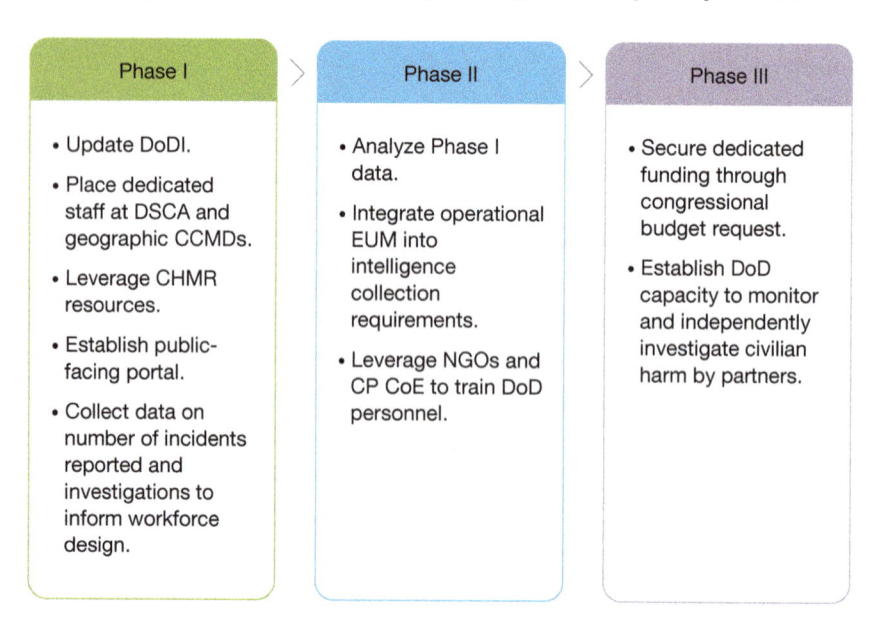

Phase I	Phase II	Phase III
• Update DoDI. • Place dedicated staff at DSCA and geographic CCMDs. • Leverage CHMR resources. • Establish public-facing portal. • Collect data on number of incidents reported and investigations to inform workforce design.	• Analyze Phase I data. • Integrate operational EUM into intelligence collection requirements. • Leverage NGOs and CP CoE to train DoD personnel.	• Secure dedicated funding through congressional budget request. • Establish DoD capacity to monitor and independently investigate civilian harm by partners.

stand gaps, challenges, and requirements in implementing the program. This pilot could allow OSD to establish a preliminary threshold and associated requirements for triggering an investigation. Information gathered through the proposed ATAP and a centralized incident reporting system (discussed in more detail below) should be used to inform policy decisions on arms transfers. Policy changes could include reconsideration of pending arms transfers or harm remediation requirements. To facilitate policy accountability, the interagency should consider establishing thresholds that trigger automatic policy reviews, such as repeated violations by a single partner or violations that cause harm of a certain magnitude. When these thresholds have been agreed upon, the interagency should establish a process by which these triggers are identified by the implementers of the proposed ATAP and rapidly communicated to the relevant interagency policymakers.

After investigations are conducted, OSD can begin to gather information on the full breadth of skills required to undertake them and determine what resources would be required for implementing the newly scoped EUM program. OSD should also collect data on the number of incidents reported and the number of investigations triggered to get a sense for the volume of staff and resources that might be required to fully implement the new DoDI.[7]

In Phase II, OSD should begin shaping skills and expertise among existing personnel in DoD to support DoDI implementation. To start, OSD should analyze data collected during Phase I to identify personnel requirements, recognizing that the size of programs could vary across regions and time depending on the volume of transfers and the presence of conflict. OSD should also consult with the intelligence community to analyze insights from Phase I to inform new intelligence collection requirements to support the EUM process and address any persistent and recurring gaps that inhibit DoD's ability to make conclusive determinations.

During Phase II, OSD might also benefit from leveraging subject-matter expertise from the CP CoE and some external expertise to begin training staff on competencies needed to implement the new DoDI. There are several organizations in the civilian harm mitigation community of interest that have significant experience investigating civilian harm resulting from military operations.[8] OSD might consider working with those organizations in addition to experts in DoD and the U.S. interagency.

Finally, in addition to training existing staff, DoD will likely need to hire new staff in Phase III to adopt a global scope for Golden Sentry and to establish an independent capacity to monitor and investigate civilian harm

---

[7] Analysis and findings from Phase I should be considered in conjunction with findings of the ongoing CHMR-AP manpower study, a DoD-wide workforce study being undertaken by the U.S. Army Manpower Analysis Agency in coordination with other DoD components to determine personnel needs for the department's implementation of the CHMR-AP, which will assess whether the human capital investments initially requested will be sufficient for implementation. The findings of this manpower study might yield relevant insights to inform resourcing an expanded EUM program. For more information on the study, see Diana Maurer, *Civilian Harm: DOD Should Take Actions to Enhance Its Plan for Mitigation and Response Efforts*, U.S. Government Accountability Office, GAO-24-106257, March 14, 2024.

[8] Organizations, such as Amnesty International and Bellingcat, have published investigations of harm caused by partners' use of U.S.-origin equipment. See, for example, Amnesty International, 2023b, and Aric Toler, "American-Made Bomb Used in Airstrike on Yemen Wedding," Bellingcat, April 27, 2018.

stemming from partners' use of U.S.-origin equipment. The right structure and organization of staffing for the proposed ATAP will depend on OSD's findings in Phases I and II, and OSD should leverage data collected during those phases to inform budget requests and staffing decisions. However, we expect that implementing the proposed ATAP will require a limited presence of permanent staff at each CCMD for routine monitoring, with a larger roving team available for reach-back support for investigations, as needed. We anticipate that this ad hoc staffing model will be an efficient use of resources, given the variable requirements across time and space for investigative support.

Expanding the scope of the EUM process will require a strategic and well-resourced approach. By taking a phased approach to evaluating, requesting, and integrating new resources and personnel, OSD can facilitate smooth implementation of a new DoDI. The phased approach—starting with data collection and pilot implementation, followed by skill development and training, and culminating in the hiring of new staff—ensures that OSD can adaptively respond to the emerging needs of the newly scoped EUM process.

Collaborate Across the Interagency to Establish a Centralized System for Collecting Public Reports of Harm by Partners Using U.S.-Origin Defense Articles

To maintain a robust system that monitors U.S. partners' use of U.S.-origin equipment, the U.S. government should be sure that the external voices—including affected populations, NGOs, and civil society groups—can easily report instances of civilian harm caused by partners operating with U.S.-origin equipment. DoD maintains a process by which the public can report harms caused by U.S. military operations, but there is no known process by which the public can submit reports of harm stemming from partners' use of U.S.-origin equipment.[9] DoS has also indicated that it plans to track harms by partners and allies, including through reporting by media and

---

[9] Blaha, 2024.

civil society groups, by way of its new CHIRG process, but DoS does not yet maintain a vehicle for members of the public to directly submit reports of civilian harm resulting from U.S.-origin weapons.[10]

Because there are the challenges of identifying cases of civilian harm resulting from U.S.-origin weapons in regions with limited U.S. military and diplomatic presence, it is clear that a publicly accessible, formal mechanism for the reporting of these cases will increase their visibility to U.S. officials and generate more meaningful and complete data regarding the scale of the issue to inform future resourcing decisions. Human Rights Reporting Gateway, a DoS website created to comply with the Leahy Act, represents a good example of how this could work, having been launched in September 2022 by DoS's Bureau of Democracy, Human Rights, and Labor—one of the offices also involved in the implementation of the CHIRG.[11] A similar approach could be developed for ATAP.

Instead of creating its own public reporting system separate from that used for the CHIRG process, OSD should work with DoS to shape one centralized intake point for all public reports of harm to be distributed across the interagency for review and action, as appropriate. One DoD representative expressed concern to us that the U.S. government could make a "science project" of reporting incidents if the interagency does not manage to coordinate the intake and distribution of relevant information.[12] Two separate processes could create confusion for those trying to submit reports and silo information in the U.S. government. To facilitate the smooth collection and distribution of reports, it will be critical for the U.S. government to adopt a coordinated approach. Ideally, DoD and DoS would create a single, public-facing portal to collect these reports.

Although DoD does not maintain a formal process for collecting public reports of harm stemming from partners' use of U.S.-origin equipment,

---

[10] CHIRG is referenced as a means "to enhance civilian protection wherever U.S.-origin defense articles are employed" in Jessica Lewis, "The Future of Security Sector Governance," U.S. Department of State, Bureau of Political-Military Affairs, February 13, 2024; also see Blaha, 2024.

[11] Human Rights Reporting Gateway, homepage, U.S. Department of State, undated.

[12] U.S. government expert on civilian harm mitigation and security assistance, interview with the authors, February 9, 2024.

DoD has long been investigating reports of civilian harm resulting from its own operations. OSD and the interagency can leverage institutional knowledge on the successes, gaps, and challenges of these programs to develop a new system for reporting and investigating partners' harms that facilitates the smooth flow of information.

Define Consistent Standards for Evidence in Investigations to Facilitate Intake of Civilian Harm Reports from External Sources

For a centralized intake system to be useful and accessible, DoD must work with the interagency to jointly define clear and consistent standards for evidence when investigating reports of harm by ally or partner forces—and then clearly communicate those standards to the public.

When developing and communicating evidentiary standards, DoD should incorporate lessons learned from its past efforts to incorporate public reporting into investigations of civilian harm caused by its own forces. DoD acknowledged in its own 2018 *Civilian Casualty (CIVCAS) Review* that it *should* "systematically seek out additional sources on potential civilian casualties" and engage more closely with NGOs in the reporting and investigation of incidents.[13] However, even though NGOs are eager to partner with DoD, they have encountered difficulty in understanding the poorly defined language that DoD uses in its civilian casualty assessments.[14] Furthermore, NGOs have noted changes in DoD's standards for evidence—such as new requirements that the multimedia submitted be accompanied by metadata—without public notification of those changes.[15] If DoD wants to harness the potential of civil society to understand instances of civilian harm caused by partners operating with U.S.-origin defense articles, the department must provide clear guidelines to the public on the kinds of information that it accepts as credible when assessing reports of harm. It is

---

[13] Joint Chiefs of Staff, *Civilian Casualty (CIVCAS) Review*, April 17, 2018, p. 3.

[14] Airwars, *US Military Assessments of Civilian Harm: Lessons Learned from the International Fight Against ISIS*, March 2019

[15] Emily Tripp, Anna Zahn, Paras Shah, Tiffany Chang, Michelle Eigenheer, and Clara Apt, "A New Standard for Evidence of Civilian Harm?" *Just Security* podcast, August 23, 2023.

also important to recognize the risks of availability bias when relying on external reports. Facilitating the intake of external reports and strengthening DoD's ability to collect and investigate can help mitigate the risk of gaps and blind spots.

Establishing a centralized intake system for public reports of harm and defining clear, consistent standards for evidence are essential steps for the U.S. government to effectively monitor the use of U.S.-origin equipment by its partners and will be important in shaping monitoring policies and determining a more accurate picture of required resources. By collaborating with interagency partners, DoD can ensure a coordinated approach that prevents information silos and facilitates efficient reporting by external partners. Clear communication of evidence standards to the public will also empower civil society to contribute valuable information, thereby strengthening the U.S. government's ability to understand and mitigate civilian harm caused by partners' use of U.S.-origin defense articles.

Establish an External Advisory Board for Operational End-Use Monitoring Investigations

While the U.S. government builds its own investigative capacity, it should consider the unique value that external partners can provide and formally incorporate their expertise through an external advisory board. External partners, such as NGOs, media, and civil society organizations, have experience conducting these kinds of investigations and can advise U.S. officials on how to build and refine their own investigative capacity. Additionally, given their on-the-ground presence and long-standing relationships with civilian populations, these external organizations often have access to information and networks that the U.S. government does not. The U.S. government should thus also leverage the board on an ad hoc basis to advise on where to find useful and credible sources of information for specific investigations.

Develop a New Designation System for Weapon Transfers According to Their Risks of Causing Civilian Harm

The United States transfers a large volume of defense equipment and weapons to its partners. Sales of U.S. military equipment to foreign governments reached a new record of $238 billion in 2023.[16] Monitoring partners' operational use of *all* transferred defense articles on a regular basis to track civilian harm would likely constitute an unreasonable and inefficient use of resources. Instead of attempting to monitor partners' operational use of every article transferred, OSD should calibrate the cadence and nature of monitoring to the overall level of risk associated with different transfers. To start, OSD should develop a new designation system for monitoring transferred defense articles according to the weapon itself, the supported partner, and the context in which the weapon is used.

Chapter 8 of the SAMM describes an interagency designation process for assigning defense articles monitored under Golden Sentry as for either REUM or EEUM.[17] This designation process works well for ensuring that resources are allocated efficiently across defense articles of varying degrees of risk. However, this process assesses risk on the basis of export policy and technology security concerns. Weapons that have been known to cause great harm to civilian populations and civilian infrastructure—such as white phosphorus and cluster munitions—are designated for REUM, given that the designation process is concerned mainly with technology security. DoD should continue using this designation system for implementing Golden Sentry, but OSD should work with DSCA and the interagency civilian harm mitigation community to develop a new designation system for implementing the proposed ATAP that uses a new risk paradigm that assesses the risk that weapons will pose to civilian populations and environments.

---

[16] Mike Stone, "US Arms Exports Hit Record High in Fiscal 2023," Reuters, January 29, 2024.

[17] DSCA, 2024a.

Consider Weapon Type, Operational Context, and Partner's Civilian Harm Mitigation Will and Capability When Measuring Risk

There are several dimensions that DoD should consider when categorizing the risk associated with transfers, including the inherent risk posed by the weapon, the will of the partner to minimize civilian harm, the capability of the partner to employ the weapon as intended, and the operational context of the recipient country. Notably, several of these factors are time-variant, so DoD should routinely update its assessments of overall risk according to changes in these factors over time. By considering a combination of these factors, DoD can establish a risk hierarchy for its transferred weapons and calibrate monitoring resources accordingly. For transfers of the greatest risk, OSD should consider increasing the cadence of monitoring engagements and the level of effort that U.S. personnel dedicate to understanding partners' operational use.

Weapons pose varying levels of risk to civilians from their inherent technological features and the scale of their impacts. When assessing risks associated with a weapon system, DoD should consider the weapon's ability to discriminate and its ability to contain impact. For example, explosive weapons, such as bombs, missiles, and artillery shells, have wide-area effects that can harm nearby civilians, even when directed at military targets.[18] OSD should review the department's data on civilian casualties—in addition to consulting civilian harm mitigation experts—to understand which weapons have the most potential for causing civilian harm.

Additionally, although the design of weapon systems can affect the potential for civilian harm, the operational context in which they are used might also shape risk to civilians. For this reason, we recommend that DoD considers the recipient government's operational context when categorizing transfers for monitoring. The presence of conflict should be the first consideration; whether a partner government is actively engaged in conflict could have a significant effect on risk levels. Additionally, the character of conflict matters in this context. To build on the example above, explosive

---

[18] Eirini Giorgou, *Explosive Weapons with Wide Area Effects: A Deadly Choice in Populated Areas*, International Committee of the Red Cross, January 2022.

weapons pose much higher risks to civilians when used in densely populated areas than less densely populated ones.[19] In addition to the geography of the conflict, DoD should consider the overall pace and intensity of the conflict when categorizing transfer risk. Fast-paced and high-intensity conflicts might have a greater potential for causing harms on a large scale. Even though this will function as a policy designation, as opposed to a legal one, principles contained in DoD's *Law of War Manual,* including "inherently indiscriminate weapons" and "weapons causing superfluous injury," might be useful concepts in the development of this risk assessment.[20] These assessments and the classification of weapons that follow should be informed by data regarding patterns of incidental civilian harm according to weapon effects and shaped by population dynamics, civilian infrastructure, and terrain.[21]

Partners' will and capability for employing weapons in a manner that reduces civilian harm are also important dimensions to consider. To measure partners' will, DoD should consider using findings from the CBAPs tasked to DSCA by the CHMR-AP and the findings from investigations triggered through the proposed ATAP to measure partner will and capacity to mitigate civilian harm. Even if a partner intends to avoid harming civilians, that partner might demonstrate a limited capability for doing so. An NGO representative explained in an interview that, despite long histories of receiving U.S. security assistance, some partners' militaries still lack sufficient technical expertise to drop munitions accurately on their intended targets.[22] A lack of sufficient intelligence can also preclude militaries from employing even the most precise weapons with accuracy.

---

[19] See EWIPA, 2022, and DoS, Office of the Spokesperson, "United States Endorses Political Declaration Relating to Protection of Civilians in Armed Conflict," press release, November 18, 2022.

[20] See DoD, 2023a.

[21] Sahr Muhammedally and Daniel Mahanty, "The Human Factor: The Enduring Relevance of Protecting Civilians in Future Wars," *Texas National Security Review,* Vol. 5, No. 3, Summer 2022.

[22] NGO expert on civilian harm investigation, interview with the authors, January 25, 2024.

To align ATAP resources appropriately, DoD should consider taking an enhanced approach to monitoring the use of weapons by partners with a poor track record (and thus greater risk) for the protection of civilians during conflict, whether because of a lack of partner capability or will. DoD could also consider taking an approach more in line with the "trust but verify" principle for partners that have demonstrated consistency with respect to their will and also possess highly developed capabilities for employing weapons in a manner that minimizes civilian harm.

In conclusion, OSD should adopt a risk-based approach to monitoring the operational use of transferred defense articles with respect to civilian harm. By developing a new designation system for the proposed ATAP that considers risks to civilians, DoD can allocate resources more efficiently. By routinely updating risk assessments according to these factors, monitoring efforts can be made more targeted and adaptive. As operational monitoring gradually yields better data on the types of weapons and contexts most strongly associated with civilian harm, these data should be fed back into risk assessments. By leveraging the relevant data and expertise, DoD can implement a risk-based framework for the proposed ATAP, focusing on monitoring activities that are designed to mitigate the risks of arms transfers to civilians, while Golden Sentry continues to monitor for technological security concerns.

Concluding Thoughts

An expanded scope for EUM will ultimately lead to a program that looks quite different from existing approaches. The scope and processes designed for Golden Sentry are established for a specific purpose: to mitigate the risks of diversion and the threat it poses to U.S. technological security. Mitigating risks of civilian harm will rely on a more comprehensive approach to monitoring to generate necessary visibility into operational use. This type of monitoring will be more labor-intensive *and* will require the contributions of a wider variety of skills from DoD, DoS, and NGOs.

But all EUM efforts, both existing and future ones, respond to a common strategic imperative—the need to mitigate the risks that are created in the course of U.S. arms transfers. The prevailing threat consideration when

Golden Sentry EUM was first established was the retention of U.S. technological supremacy, but today's policy context includes a growing concern that the U.S. commitment to advancing human rights be reflected in U.S. arms transfers. Similar to how Golden Sentry helps to mitigate against the risks of diversion, a new program that monitors partner operational use of transferred arms can help to guard against the risks that transferred U.S.-origin arms undermine U.S. interests through unlawful use or operational employment that is inconsistent with U.S. values. Use of U.S.-origin defense articles by partners in ways that violate U.S. norms, laws, and values can threaten U.S. interests by undermining the United States' global reputation and casting doubt on U.S. commitments to improve the protection of civilians in conflict. In some cases, these instances will contribute to increased instability and the escalation of conflict, which will implicate the United States because of the arms-transfer relationship.

The principles embedded in the existing regimes for governing weapon transfer and use provide a helpful starting point for the expansion of EUM, and recent DoD investments in civilian harm mitigation and reduction capabilities will yield new expertise relevant to this mission over time. Efforts to mitigate civilian harm caused by U.S. operations and the approaches DoD adopts to mitigate the risks inherent in arms transfers should be integrated to fullest degree possible. There is significant potential for efficiencies through sharing expertise. Conceptualizing operational EUM as a part of an overarching approach to weapon life-cycle management might prove to be a helpful way to build in the required costs to enhance U.S. oversight, if the U.S. arms sales' "total package approach is inclusive of mitigating civilian harm,"[23] as officials have stated in the past. Ultimately, to reduce civilian harm resulting from partners' use of U.S.-supplied weapons, operational monitoring must feed into a process by which the policies regarding arms transfers are reconsidered in light of the findings. This need not always be punitive nor must it necessarily result in a suspension of arms transfers. In

---

[23] DoD, "DOS and DOD Officials Brief Reporters on Fiscal 2020 Arms Transfer Figures," transcript of remarks by R. Clark Cooper and Heidi Grant, December 4, 2020.

some cases, it might indicate that capability gaps can be addressed through training and improvements to equipment.[24]

DoD has committed to hold itself to a higher standard than ever before in its efforts to mitigate civilian harm. Raising the bar in the standards applied in security cooperation programs, especially in the transfer of lethal weapons, and developing robust mechanisms to ensure partner compliance are consistent with that commitment.

---

[24] DoD and DoS established the Targeting Working Group in June 2019 to reduce the risk of partner nation or coalition operations causing civilian harm. The Targeting Working Group was a joint, interagency effort co-led by the Bureau of Political-Military Affairs and DSCA to ensure that appropriate targeting infrastructure capabilities for certain munitions, notably precision-guided munitions, are accounted for whenever such munitions are sold or transferred via FMS or other U.S. government security cooperation activities.

Abbreviations

AECA	Arms Export Control Act
ATAP	Arms Transfer Accountability Program
CAT	Conventional Arms Transfer
CAV	Compliance Assessment Visit
CBAP	Civilian Harm Mitigation and Response Baselines of Allies and Partners
CCMD	combatant command
CHAC	civilian harm assessment cell
CHIRG	Civilian Harm Incident Response Guidance
CHMR-AP	Civilian Harm Mitigation and Response Action Plan
CP CoE	Civilian Protection Center of Excellence
DoD	U.S. Department of Defense
DoDI	Department of Defense instruction
DoS	U.S. Department of State
DSCA	Defense Security Cooperation Agency
EEUM	enhanced end-use monitoring
EUM	end-use monitoring
EWIPA	Political Declaration on Strengthening the Protection of Civilians from the Humanitarian Consequences Arising from the Use of Explosive Weapons in Populated Areas
FMF	foreign military financing
FMS	foreign military sales
FY	fiscal year
GAO	U.S. Government Accountability Office
IHL	international humanitarian law
JAG	judge advocate general
NGO	nongovernmental organization
NSM-20	National Security Memorandum 20

OSD	Office of the Secretary of Defense
PM	Bureau of the Political-Military Affairs
REUM	routine end-use monitoring
SAMM	*Security Assistance Management Manual*
SCIP	Security Cooperation Information Portal
SCO	Security Cooperation Organization
VCA	Virtual Compliance Assessment

References

Abdelaziz, Salma, Alla Eshchenko, and Joe Sterling, "Saudi-Led Coalition Admits 'Mistakes' Made in Deadly Bus Attack in Yemen," CNN, September 2, 2018.

Airwars, *US Military Assessments of Civilian Harm: Lessons Learned from the International Fight Against ISIS*, March 2019.

Airwars, Amnesty International USA, Anera, et al., civilian harm in Gaza, letter to Lloyd J. Austin III, December 20, 2023. As of November 18, 2024: https://civiliansinconflict.org/wp-content/uploads/2023/12/NGO-Letter-to-Secretary-Austin-on-Civilian-Harm-in-Gaza.pdf

Amnesty International, "Lebanon: Evidence of Israel's Unlawful Use of White Phosphorus in Southern Lebanon as Cross-Border Hostilities Escalate," October 31, 2023a.

Amnesty International, "Israel/OPT: US-Made Munitions Killed 43 Civilians in Two Documented Israeli Air Strikes in Gaza—New Investigation," December 5, 2023b.

Armed Conflict Location and Event Data, "Country Hub: Yemen," webpage, undated. As of October 30, 2024: https://acleddata.com/middle-east/yemen/

Bagshaw, Simon, "The 2022 Political Declaration on the Use of Explosive Weapons in Populated Areas: A Tool for Protecting the Environment in Armed Conflict?" *International Review of the Red Cross*, No. 954, December 2023.

Blaha, Charles O. (Cob), "The State Department's Civilian Harm Incident Response Guidance: How to Make a Good Thing Better," *Just Security* blog, July 18, 2024.

Blair, Jason L., *Yemen: State and DOD Need Better Information on Civilian Impacts of U.S. Military Support to Saudi Arabia and the United Arab Emirates*, U.S. Government Accountability Office, GAO-22-105988, June 2022.

Chappell, John Ramming, and Ari Tolany, "Unpacking Biden's Conventional Arms Transfer Policy," Lawfare Institute and Brookings Institution, March 1, 2023.

Clermont, Kevin M., "Staying Faithful to the Standards of Proof," *Cornell Law Review*, Vol. 104, No. 6, September 2019.

Coble, Kevin S., and John C. Tramazzo, "Israel-Hamas 2023 Symposium—White Phosphorus and International Law," *Articles of War* blog, October 25, 2023.

Conflict Armament Research, "Typology of Diversion: A Statistical Analysis of Weapon Diversion Documented by Conflict Armament Research," *Diversion Digest*, No. 1, August 2018.

Cresswell, John W., and Vicki L. Plano Clark, *Designing and Conducting Mixed Method Research*, 2nd ed., SAGE Publications, 2010.

Cruickshank, Michael, "A Saudi War-Crime in Yemen? Analysing the Dahyan Bombing," Bellingcat, August 18, 2018.

C-SPAN, "Arms Control Association Discussion on U.S. Arms Transfers," video, June 7, 2024. As of August 27, 2024: https://www.c-span.org/video/?536232-5/arms-control-association-discussion-us-arms-transfers

Davis, Christopher M., *The Amending Process in the Senate*, Congressional Research Service, 98-853, September 16, 2015.

Defense Security Cooperation Agency, *FY 2010 Congressional Budget Justification: Foreign Assistance, Title IV Supporting Information*, 2009.

Defense Security Cooperation Agency, "End Use Monitoring (EUM) Responsibilities in Support of the Department of Defense Golden Sentry EUM Program," DSCA Policy Memorandum No. 02-43, December 4, 2002.

Defense Security Cooperation Agency, "DSCA DoD Golden Sentry End-Use Monitoring (EUM) Virtual Compliance Assessments (VCA)," DSCA Policy Memorandum No. 21-64, October 13, 2021.

Defense Security Cooperation Agency, "Conducting End Use Monitoring in a Hostile Environment," DSCA Policy Memorandum No. 22-87, December 20, 2022.

Defense Security Cooperation Agency, "FY22 Security Cooperation Figures Announcement," press release, January 25, 2023.

Defense Security Cooperation Agency, "End Use Monitoring," in *Security Assistance Management Manual*, 2024a.

Defense Security Cooperation Agency, "LOA Notes Listing," in *Security Assistance Management Manual*, 2024b.

Defense Security Cooperation Agency, "Security Assistance Forecasting, Planning, Programming, Budgeting, Execution, and Audit," in *Security Assistance Management Manual*, 2024c.

Defense Security Cooperation University, "End-Use Monitoring and Third-Party Transfers," in *Security Cooperation Management*, 43rd ed., 2023a.

Defense Security Cooperation University, "Security Cooperation Organizations Overseas," in *Security Cooperation Management*, 43rd ed., 2023b.

References

Dent, Elizabeth, and Grant Rumley, "How the U.S. Used Arms Sales to Shift Saudi Behavior," *Defense One*, September 4, 2024.

DoD—*See* U.S. Department of Defense.

DoS—*See* U.S. Department of State.

DSCA—*See* Defense Security Cooperation Agency.

EWIPA—*See* Political Declaration on Strengthening the Protection of Civilians from the Humanitarian Consequences Arising from the Use of Explosive Weapons in Populated Areas.

Forum on the Arms Trade, "Moving Beyond Rhetoric: Overcoming Challenges to Better U.S. Policy," video, June 4, 2024. As of October 15, 2024: https://www.youtube.com/live/AWSa9UV3TcI

Giorgou, Eirini, *Explosive Weapons with Wide Area Effects: A Deadly Choice in Populated Areas*, International Committee of the Red Cross, January 2022.

Human Rights Reporting Gateway, homepage, U.S. Department of State, undated. As of November 26, 2024: https://hrgshr.state.gov/

Huntley, Todd, "Airstrikes, Civilian Casualties, and the Role of JAGs in the Targeting Process," Lawfare Institute and Brookings Institution, August 22, 2022.

Joint Chiefs of Staff, *No-Strike and the Collateral Damage Estimation Methodology*, Chairman of the Joint Chiefs of Staff Instruction 3160.01A, October 12, 2012.

Joint Chiefs of Staff, *Civilian Casualty (CIVCAS) Review*, April 17, 2018.

Joint Chiefs of Staff, "Enclosure C—Joint Manpower and Program: Requirements Determination and Validation," in *Joint Manpower and Personnel Program*, Chairman of the Joint Chiefs of Staff Instruction 1001.01C 21, February 21, 2024.

Kenney, Chelsa A., *Ukraine: DOD Should Improve Data for Both Defense Article Delivery and End-Use Monitoring*, U.S. Government Accountability Office, GAO-24-106289, March 13, 2024.

Kerr, Paul K., and Michael A. Weber, *U.S. Arms Sales and Human Rights: Legislative Basis and Frequently Asked Questions*, Congressional Research Service, IF11197, February 29, 2024.

Khan, Azmat, "The Human Toll of America's Air Wars," *New York Times*, December 19, 2021.

Krause, Keith, "Military Statecraft: Power and Influence in Soviet and American Arms Transfer Relationships," *International Studies Quarterly*, Vol. 35, No. 3, September 1991.

Kurç, Çağlar, "No Strings Attached: Understanding Turkey's Arms Exports to Africa," *Journal of Balkan and Near Eastern Studies*, Vol. 26, No. 3, 2024.

Lewis, Jessica, "The Future of Security Sector Governance," U.S. Department of State, Bureau of Political-Military Affairs, February 13, 2024.

Mahanty, Daniel, and Annie Shiel, *With Great Power: Modifying US Arms Sales to Reduce Civilian Harm,* Center for Civilians in Conflict and Stimson Center, 2018.

Maurer, Diana, *Civilian Harm: DOD Should Take Actions to Enhance Its Plan for Mitigation and Response Efforts*, U.S. Government Accountability Office, GAO-24-106257, March 14, 2024.

McNerney, Michael J., Jonah Blank, Becca Wasser, Jeremy Boback, and Alexander Stephenson, *Improving Implementation of the Department of Defense Leahy Law*, RAND Corporation, RR-1737-OSD, 2017. As of October 9, 2024:
https://www.rand.org/pubs/research_reports/RR1737.html

McNerney, Michael J., Gabrielle Tarini, Karen M. Sudkamp, Larry Lewis, Michelle Grisé, and Pauline Moore, *U.S. Department of Defense Civilian Casualty Policies and Procedures: An Independent Assessment*, RAND Corporation, RR-A418-1, 2022. As of August 19, 2024:
https://www.rand.org/pubs/research_reports/RRA418-1.html

Muhammedally, Sahr, and Daniel Mahanty, "The Human Factor: The Enduring Relevance of Protecting Civilians in Future Wars," *Texas National Security Review*, Vol. 5, No. 3, Summer 2022.

Office of U.S. Senator Patty Murray, "Values in Arms Export Act of 2022," February 2022. As of October 14, 2024:
https://www.murray.senate.gov/wp-content/uploads/2022/02/VAEA-One-Pager.pdf

New America, "The War in Yemen," webpage, undated. As of June 28, 2024:
https://www.newamerica.org/future-security/reports/americas-counterterrorism-wars/the-war-in-yemen/

Paul, Josh, "This Is Not the State Department I Know. That's Why I Left My Job," *Washington Post*, October 23, 2023.

Paul, T. V., "Influence Through Arms Transfers: Lessons from the U.S.-Pakistani Relationship," *Asian Survey*, Vol. 32, No. 12, December 1992.

Pierre, Andrew J., "Arms Sales: The New Diplomacy," *Foreign Affairs*, Vol. 60, No. 1, Winter 1981/1982.

Political Declaration on Strengthening the Protection of Civilians from the Humanitarian Consequences Arising from the Use of Explosive Weapons in Populated Areas (EWIPA), signed at Dublin, Ireland, November 18, 2022.

Protocol on Prohibitions or Restrictions on the Use of Incendiary Weapons (Protocol III), adopted at Geneva, Switzerland, October 10, 1980.

Public Law 87-195, Foreign Assistance Act of 1961, as amended through Public Law 118-83, September 26, 2024.

Public Law 90-629, Arms Export Control Act, as amended through Public Law 118-31, National Defense Authorization Act for Fiscal Year 2024, December 22, 2023.

Rajput, Aniruddha, "Standard of Proof," in Anne Peters, ed., *Max Planck Encyclopedias of International Law*, Oxford University Press, February 2021.

Ryan, Missy, "Biden Administration Will Track Civilian Deaths from U.S.-Supplied Arms," *Washington Post*, September 13, 2023.

Schmitt, Michael N., "Military Necessity and Humanity in International Humanitarian Law: Preserving the Delicate Balance," *Virginia Journal of International Law*, Vol. 50, No. 4, May 2010.

Sharp, Jeremy M., *Yemen: Civil War and Regional Intervention*, Congressional Research Service, R43960, November 23, 2021.

Shiel, Annie, Seth Binder, Jeff Abramson, William Hartung, Rachel Stohl, Diana Ohlbaum, Adam Isacson, Brittany Benowitz, and Daniel R. Mahanty, "Toward a More Responsible US Arms Trade Policy: Recommendations for the Biden-Harris Administration," *Just Security* blog, January 19, 2021.

Shostak, Melissa, and Cortney Weinbaum, "How China Is Building Influence Through Arms Sales," *National Interest*, December 7, 2022.

Stigall, Dan, "The EWIPA Declaration and U.S. Efforts to Minimize Civilian Harm," *Articles of War* blog, May 22, 2024.

Stockholm International Peace Research Institute, "Trends in International Arms Transfers, 2023," fact sheet, March 2024.

Stone, Mike, "US Arms Exports Hit Record High in Fiscal 2023," Reuters, January 29, 2024.

Thrall, A. Trevor, Jordan Cohen, and Caroline Dorminey, "Power, Profit, or Prudence? US Arms Sales Since 9/11," *Strategic Studies Quarterly*, Vol. 14, No. 2, Summer 2020.

Tolany, Ari, Alexander Bertschi Wrigley, Annie Shiel, Elias Yousif, and Lauren Woods, "Demystifying End-Use Monitoring in U.S. Arms Exports," Center for Civilians in Conflict, Security Assistance Monitor, and Stimson Center, September 16, 2021.

Toler, Aric, "American-Made Bomb Used in Airstrike on Yemen Wedding," Bellingcat, April 27, 2018.

Tripp, Emily, Anna Zahn, Paras Shah, Tiffany Chang, Michelle Eigenheer, and Clara Apt, "A New Standard for Evidence of Civilian Harm?" *Just Security* podcast, August 23, 2023.

U.S. Central Command, "Senate Armed Services Committee Hearing Posture of United States Central Command and United States Africa Command in Review of the Defense Authorization Request for Fiscal Year 2025 and the Future Years," transcript, March 8, 2024.

U.S. Code, Title 10, Chapter 16, Section 384, Department of Defense Security Cooperation Workforce Development.

U.S. Code, Title 22, Chapter 31, Section 2304, Human Rights and Security Assistance.

U.S. Code, Title 22, Chapter 39, Section 2785, End-Use Monitoring of Defense Articles and Defense Services.

U.S. Department of Defense, *Registration and End-Use Monitoring of Defense Articles and/or Defense Services*, Department of Defense Instruction 4140.66, incorporating change 1, May 24, 2017.

U.S. Department of Defense, "DOS and DOD Officials Brief Reporters on Fiscal 2020 Arms Transfer Figures," transcript of remarks by R. Clark Cooper and Heidi Grant, December 4, 2020. As of April 25, 2024: https://www.defense.gov/News/Transcripts/Transcript/Article/2437110/dos-and-dod-officials-brief-reporters-on-fiscal-2020-arms-transfer-figures/

U.S. Department of Defense, *End-Use Monitoring of Defense Articles and Services Government-to-Government Services*, 2021.

U.S. Department of Defense, *End-Use Monitoring of Defense Articles and Services Government-to-Government Services*, 2022a.

U.S. Department of Defense, *Civilian Harm Mitigation and Response Action Plan (CHMR-AP)*, August 25, 2022b.

U.S. Department of Defense, *Law of War Manual*, updated July 2023a.

U.S. Department of Defense, *Civilian Harm Mitigation and Response*, Department of Defense Instruction 3000.17, December 21, 2023b.

U.S. Department of Defense, "DOD Announces Release of Department of Defense Instruction and Website on Civilian Harm Mitigation and Response," press release, December 21, 2023c.

U.S. Department of Defense, "Fiscal Year 2023 U.S. Arms Transfers and Defense Trade," fact sheet, January 29, 2024.

U.S. Department of Defense, Office of the Inspector General, "Management Advisory: DoD Review and Update of Defense Articles Requiring Enhanced End-Use Monitoring," DODIG-2023-074, May 19, 2023.

U.S. Department of State, "About the Leahy Law," fact sheet, Bureau of Democracy, Human Rights, and Labor, January 20, 2021a.

U.S. Department of State, "End-Use Monitoring of U.S.-Origin Defense Articles," fact sheet, Bureau of Political-Military Affairs, January 20, 2021b.

U.S. Department of State, "U.S. Security Cooperation with Ukraine," fact sheet, Bureau of Political-Military Affairs, December 2, 2024.

U.S. Department of State, *Report to Congress Under Section 2 of the National Security Memorandum on Safeguards and Accountability with Respect to Transferred Defense Articles and Defense Services (NSM-20)*, May 10, 2024.

U.S. Department of State, Office of the Inspector General, *Review of the Department of State's Role in Arms Transfers to the Kingdom of Saudi Arabia and the United Arab Emirates*, August 2020.

U.S. Department of State, Office of the Spokesperson, "United States Endorses Political Declaration Relating to Protection of Civilians in Armed Conflict," press release, November 18, 2022.

U.S. House of Representatives, Safeguarding Human Rights in Arms Exports Act of 2023 or the SAFEGUARD Act of 2023, H.R. 1801, March 27, 2023.

U.S. Senate, Values in Arms Export Act of 2022, S. 3558, February 2, 2022.

U.S. Senate, Safeguarding Human Rights in Arms Exports Act of 2023 or the SAFEGUARD Act of 2023, S.1025, March 29, 2023.

U.S. Senate, Safeguards and Accountability with Respect to Transferred Defense Articles and Defense Services, Amendment 1389 to H.R. 815, submitted February 7, 2024.

Van Hollen, Chris, "Israel's War Against Hamas Is Just, but It Must Be Fought Justly," *Washington Post*, December 6, 2023.

Van Hollen, Chris, Brian Schatz, Martin Heinrich, et al., National Security Memorandum on Safeguards and Accountability with Respect to Transferred Defense Articles and Defense Services implementation, letter to Antony Blinken, Lloyd J. Austin III, and Avril Haines, U.S. Senate, March 8, 2024. As of November 19, 2024:
https://www.vanhollen.senate.gov/imo/media/doc/nsm_implementation_letter.pdf

Van Schaack, Beth, "Evaluating Proportionality and Long-Term Civilian Harm Under the Laws of War," *Just Security* blog, August 29, 2016.

Vergun, David, "Officials Describe How Arms Sales Benefit the U.S., Partners," U.S. Department of Defense, December 4, 2020.

Warren, Elizabeth, Jeffrey A. Merkley, Bernard Sanders, Tim Kaine, and Martin Heinrich, civilian harm from Israel's use of U.S. weapons in Gaza, letter to Joseph R. Biden, U.S. Senate, December 5, 2023. As of November 19, 2024: https://www.warren.senate.gov/imo/media/doc/2023.12.05%20Letter%20to%20President%20Biden%20re%20Civilian%20Harm%20from%20Israel%27s%20use%20of%20U.S.%20Weapons%20in%20Gaza.pdf

Warren, Elizabeth, Bernard Sanders, and Michael S. Lee, weapon use monitoring and civilian harm, letter to Joseph R. Biden, U.S. Senate, May 14, 2023. White House, "Remarks by President Biden on America's Place in the World," February 4, 2021. As of November 19, 2024: https://www.warren.senate.gov/imo/media/doc/2023.12.05%20Letter%20to%20President%20Biden%20re%20Civilian%20Harm%20from%20Israel%27s%20use%20of%20U.S.%20Weapons%20in%20Gaza.pdf

White House, "Memorandum on United States Conventional Arms Transfer Policy," National Security Memorandum, NSM-18, February 23, 2023.

White House, "National Security Memorandum on Safeguards and Accountability with Respect to Transferred Defense Articles and Defense Services," National Security Memorandum, NSM-20, February 8, 2024.

World Health Organization, "White Phosphorus," fact sheet, January 15, 2024.

Yousif, Elias, *"If We Don't Sell It, Someone Else Will": Dependence and Influence in U.S. Arms Transfers*, Stimson Center, March 2023.